D1259511

OMNIVM
LVX
CIVIVM
MDCCCLII
MDCCC
LXXVIII

DNA: THE LADDER OF LIFE

SECOND EDITION

DNA: THE LADDER OF LIFE

SECOND EDITION

BY DR. EDWARD FRANKEL

ILLUSTRATED BY RADU VERO

MCGRAW-HILL BOOK COMPANY

NEW YORK ST. LOUIS SAN FRANCISCO AUCKLAND
BOGOTÁ DÜSSELDORF JOHANNESBURG LONDON
MADRID MEXICO MONTREAL NEW DELHI PANAMA PARIS
SÃO PAULO SINGAPORE SYDNEY TOKYO TORONTO

Library of Congress Cataloging in Publication Data

Frankel, Edward.
DNA, the ladder of life.

Bibliography: p.
Includes index.
1. Deoxyribonucleic acid. I. Title.
[DNLM: 1. DNA—Popular works. QU58 F829d]
QP624.F7 1978 574.8′732 78–7879
ISBN 0–07–021883–8

123456789 BPBP 7832109

TO MY WIFE HELEN
AND OUR GRANDCHILDREN
CARA AND PETER
"DNA DELIGHTS"

PREFACE

In the history of science there are but a few instances of a complete discontinuity in the perspectives that scientists have of a particular science. For life scientists the most recent one was the evidence that genetic material is deoxyribonucleic acid (DNA). When the exquisite structure of DNA was discovered by Watson and Crick in 1953 modern biology was kicked into high gear. Dr. Frankel tells us in a remarkably clear and concise way how we arrived at the DNA revolution.

With the structure of DNA at hand two important problems were opened to study: (1) the replication of the gene (DNA), and (2) the control of gene function. Although at first it was thought that the study of gene duplication would be quite simple, only after 20 years has detailed information been obtained on the many components needed to assure faithful gene duplication.

Primary gene function was quickly understood to be determined by an RNA copy of the DNA. Within the nucleotide sequences of these RNA molecules was the genetic code for assembling in proper order the 20 amino acids that make up the many proteins that are the cell's biological catalysts and structural components. The code was deciphered by 1964.

With DNA replication understood in principle and the genetic code established, life scientists began to determine how the temporal control of gene function was brought about in bacteria; it was learned that this is accomplished by molecules that turn genes on and off in response to a variety of internal and external signals.

To study such processes in higher life forms seems an insurmountable task. The amount of genetic material involved in these forms is so large that focusing on individual genes has appeared impossible. However, just when needed, a new technique, recombinant DNA, was born. This technique will allow for the amplification and detailed study of genes from any organism. As vividly described herein, the birth of this technique has not been easy. It evoked fear in many because it allows genes to be mixed from different species, with perhaps unpredictable results. However, the scientific community itself moved to self-regulate and prescribe containment procedures for the many kinds of experiments. Others felt these procedures were inadequate and favored federal regulation of recombinant DNA experiments. As this is written the future nature of regulations is uncertain. Still, perhaps more slowly than some might wish, this technique will be used and will ultimately lead to a second revolution in molecular biology.

—Norton D. Zinder, Ph.D
Professor of Molecular Genetics
The Rockefeller University, New York, New York

ACKNOWLEDGMENTS

The content and ideas for this revised edition came from discussions with former students and current colleagues, several of whom are active DNA researchers. The raw materials were culled from innumerable scientific journals, periodicals, monographs, magazines, newspapers, and letters.

The contributions of just a handful of researchers are acknowledged by name; to the thousands not named, the author apologizes. However, the author wishes to express his thanks and gratitude to the friends and colleagues who helped in the preparation of this book. He is especially indebted to Mr. Julius Schwartz, former colleague and author of many science books for young people, for his excellent review of the original manuscript.

The entire revised manuscript was read by Dr. Norton D. Zinder of the Rockefeller University and Dr. Susan Wallace of the New York University Medical School; their critical reading and many invaluable suggestions contributed significantly in improving the text. Dr. Elena McCoy, also of the New York University Medical School, evaluated the chapter on mutations and Dr. Thomas E. Jensen of the Herbert H. Lehman College was most helpful in his critique of the chapter on the cell. My wife Helen demonstrated extraordinary patience and fortitude with the author's "highs" and "lows" during the entire writing and rewriting periods. Finally, the author alone assumes responsibility for any inaccuracies or oversights that may appear.

CONTENTS

INTRODUCTION

It is virtually impossible to keep up with the progress in any one field of science, from astronomy to zoology. Each year scientific discoveries and inventions outnumber those of the previous year. Estimates are that at the present pace scientific knowledge and technology will double in a decade. What is fantasy today may be fact tomorrow; and what we think is impossible today may be commonplace tomorrow.

Since this book was first written over a decade ago, much scientific water has flowed over the DNA dam, carrying us to new and unbelievable heights of achievement. DNA-RNA research is a mighty tributary feeding into the mainstream of science and moving with ever-increasing speed into new and unexpected fields. It is now traveling through dangerous and troubled waters that threaten to limit and impede its progress.

For anyone following the DNA-RNA drama, it may be helpful to stop for a few hours, look back over the past decade, and review what has happened so that we can understand how we got where we are and possibly where we are going. It was such considerations that strongly influenced the author to update the DNA story and share it with you.

Biochemists have moved from making Tinkertoy models of the structure of DNA in the 1950s to synthesizing the molecule in the 1970s. The original double-helix model of DNA as constructed by Francis Crick and James D. Watson provided the direction, inspiration, and incentive for making nucleic acid molecules that function in a living cell—an achievement which represents the coordinated efforts of thousands of scientists around the world. The interest and importance of this research may be judged by the fact that since 1965 almost half of the Nobel prizewinners have been so honored because of their contributions to an understanding of the nature of DNA and its chemical counterpart RNA. The discovery of how the DNA of vastly different organisms can be combined to produce genes that not even science fiction writers dreamed of has been greeted by cheers and jeers. Such fantastic feats have opened up worldwide debate and public discussion over the possible harm that could be caused should one of these DNA creations escape and start a worldwide epidemic. They think that it is possible for DNA "demons" to wreck the environment and wipe out life on earth.

It was the scientists themselves who first expressed concern about the possible damage their creations might do. And it was the scientists who called a halt to research on certain types of high-risk experiments. They set up strict rules and regulations defining the kinds of experiments that could be done, the kinds of organisms that could be used, and the laboratory conditions under which DNA research could be carried on.

It is basic to weigh the future benefits of research against its potential risks. The future of DNA research may be determined in the political arena rather than the science laboratory. In the meantime, DNA research goes on.

E. F.

DNA: THE LADDER OF LIFE

SECOND EDITION

1

DNA—THE MOLECULE OF LIFE

One of the most significant scientific advances of modern times was the discovery of a molecule that determines the basic nature of all life, from the simplest microbe to man himself. This molecule is present in every one of the billions of cells in your body. It is best known by the letters DNA, which are the initials of its chemical name—deoxyribonucleic (pronounced day-OX-ee-rye-bow-new-CLAY-ic) acid. The nucleic acid part of the name tells you that DNA is an acid located in the nucleus of the living cell.

DNA was first isolated from the nuclei of white blood cells and fish sperm cells over 100 years ago. Its importance to life, however, did not become known until a generation ago. Although the details of DNA structure are far too small to be seen with even the most powerful microscope, scientists have worked out its structure and chemical composition. They are discovering how DNA directs the processes of life in each organism. We have unlocked life's greatest secret—the genetic code—which reveals how DNA dictates the transmission of characteristics from parent to offspring. With this mystery solved, simple forms of life have been created in a test tube.

DNA, then, exercises dual control over the life of an organism. It directs metabolism, the day-to-day activities that keep the machinery of the body turning. It also determines heredity, the transmission of traits that keeps the race or species going for thousands or millions of years.

Living requires an endless flow of molecules in and out of an organism. Some of the incoming molecules are burned to provide the energy needed to power the living thing. Other molecules are transformed into the material of which the organism is composed and are used for repair, replacement, and growth. Life is an endless chain of metabolic activities and molecule manipulations which are directed and controlled by DNA.

A remarkable capacity of a living thing is its ability to reproduce, to create another very much like itself. The striking resemblance between parents and their offspring suggests that some substance is passed on from one generation to the next which determines specific traits. One of the great scientific breakthroughs of all time was the discovery that the DNA molecule is that material. Within a living cell, this molecule is able to make an exact duplicate of itself and send the copy on to direct the next generation.

THE CASE OF THE DEADLY MICROBES

Early clues about the nature of DNA came from the study of the strange behavior of certain disease-causing microbes, the pneumonia germs. There are, in general, two kinds of pneumonia bacteria, one with a jellylike coat and the other without this coat. Research showed that only the coated microbes cause pneumonia.

Back in 1928, Fred Griffith, an English bacteriologist, was experimenting with these pneumonia organisms. Using heat, he killed some of the coated germs and then injected mice with a mixture of killed, coated germs and live, harmless, uncoated microbes. From what was then known about these microbes, the mice should have remained alive and healthy. Instead, the animals developed pneumonia and died. Moreover, while the bodies of these dead mice swarmed with coated pneumonia germs, the harmless, uncoated ones could not be found. Griffith thought perhaps the killed coated germs he had injected were not really dead. He therefore injected other mice with only some dead germs from the original batch. Not one of these animals became sick. Apparently the killed germs were really dead. In some mysterious way, the uncoated microbes acquired a coat and became deadly. Evidently the dead germs passed something on to the living microbes that changed them and their offspring. What was that something?

Scientists spent the next few years puzzling over how harmless uncoated microbes could change into deadly coated ones. This mystery was finally cleared up in 1944 by Oswald T. Avery and two of his colleagues at the Rockefeller Institute in New York City. They not only isolated but also identified the

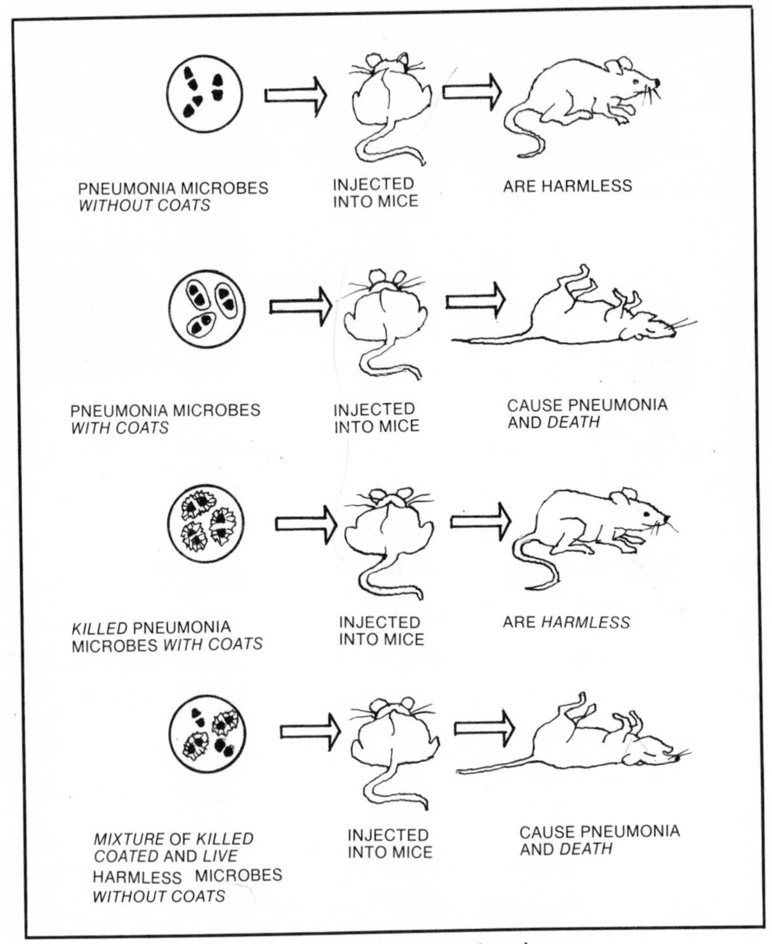

The case of the masked murdering microbes.

substance responsible for transforming these microbes. Avery extracted from the dead coated pneumonia germs a substance which could be collected as long fine threads. When harmless pneumonia microbes were grown in a solution of these threads, some of them acquired a coat and became deadly.

Equally astonishing, the change was permanent. All the offspring of the transformed microbes had coats and caused pneumonia.

Avery and his team discovered that this dramatic inherited change was due to a single chemical substance found in the long fine threads, DNA. DNA had carried the disease-causing trait from the harmful to the harmless bacteria. This experiment clearly established DNA as the specific carrier of hereditary information. Scientists were able to draw this conclusion because they observed that the DNA of one kind of organism got into another kind of organism and changed it permanently. Thus the case of the strange behavior of the dead pneumonia that seemed to come to life was solved by scientific sleuthing. The mysterious and magical transformer was DNA.

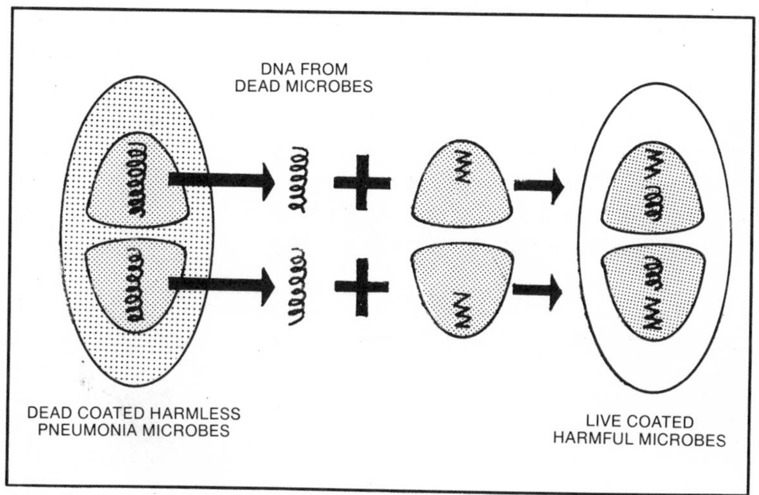

Murdering microbes unmasked.

VIRUSES—PACKAGED DNA

Another fertile source of information about DNA has come from the study of the simplest and smallest bits of living material, the viruses. They are the cause of such human diseases as measles, mumps, chickenpox, smallpox, and polio. The average virus is about one 500-thousandth of an inch long, much too small to be seen with an ordinary microscope but visible under electron-microscope magnifications.* The size of viruses places them in the twilight zone between the living and the lifeless. The smallest viruses are about the size of the largest lifeless molecules. The largest viruses are about as big as the smallest living microbes. Viruses can behave either as lifeless molecules or as living organisms. They can appear as a white powder which looks like salt crystals and display no more life than this lifeless

*Scientists use the units of measurement of the metric system, choosing units that are appropriate to the size of the object being measured. The size of bacteria is usually expressed in micrometers (MY-crow-me-ters), abbreviated by the Greek letter mu (m-you), written as μ, followed by m, thus: μm. Micro means a millionth, so a micrometer is a millionth of a meter (one 25-thousandth of an inch). The common rod-shaped bacterium that lives in the human intestines is about 2 micrometers (2 μm) long (about one ten-thousandth of an inch).

A virus is smaller than a bacterium and its size is measured in units called angstroms (symbol Å). An angstrom is one ten-thousandth of a micrometer or one ten-billionth of a meter or one 250-millionth of an inch. The diameter of an average virus is about 600 Å (one 500-thousandth of an inch). The smallest virus is 160 Å and the largest 3000 Å.

chemical. On the other hand, once a tiny virus crystal gains entry into a cell, it can come to "life" and breed new viruses at the fantastic rate of ten per

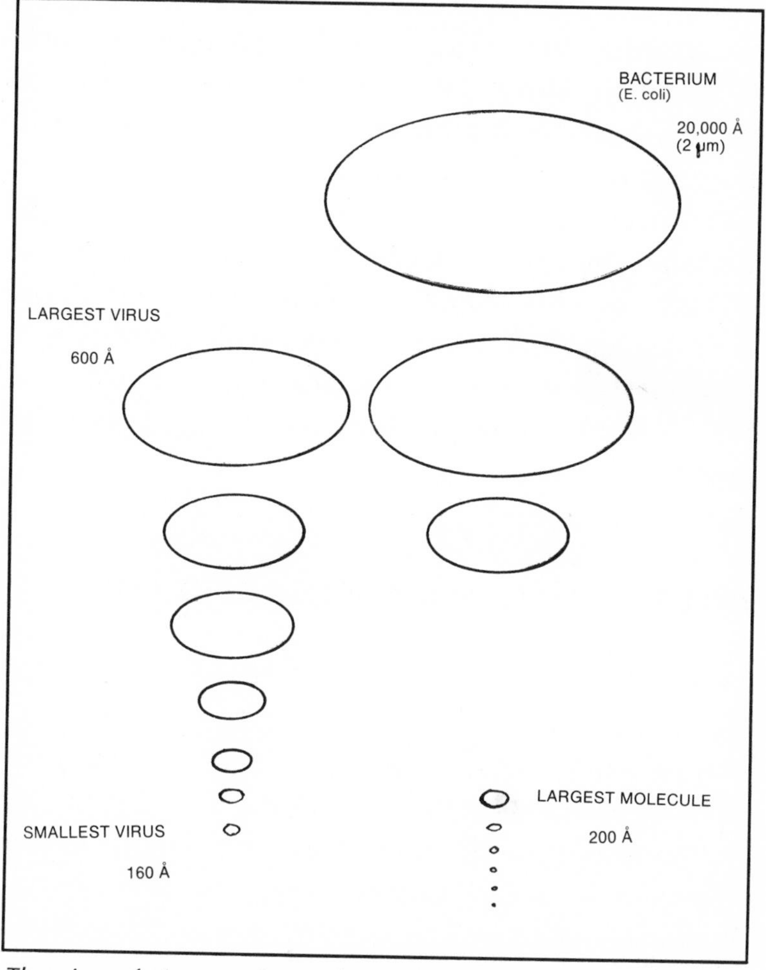

The size of viruses places them in the twilight zone between the largest lifeless molecules and the smallest living things.

second. The virus is a parasite. It can grow and reproduce only inside a living cell. The virus may be thought of as an incomplete cell. It comes to life only inside a complete cell that supplies it with the missing parts. Some viruses eat, live, and reproduce inside bacteria, which are tiny cells.

Why has the study of viruses yielded so much information about DNA? Chemically, viruses are much simpler than other living things, composed as they are only of those molecules absolutely necessary for life. They are made of just two substances, proteins and nucleic acid, welded together to make a giant molecule of nucleoproteins. The nucleic acid is coiled inside the protein coat which covers and protects it. In most viruses the nucleic acid is DNA; a few contain a related chemical, RNA, ribonucleic acid.

PROTEIN COATS AND NUCLEIC ACID CORES

In 1955 Heinz Fraenkel-Conrat, a biochemist in the Virus Laboratory at the University of California, conducted a crucial experiment that shed light on DNA as the basic molecule of life. He was studying the tobacco mosaic virus, TMV, which infects the tobacco plant and causes sickly yellow spots to appear on the leaves where the viruses have eaten away the tissues. Because of the spots, this sickness of tobacco plants is known as tobacco mosaic disease. The virus which causes this disease is rod-

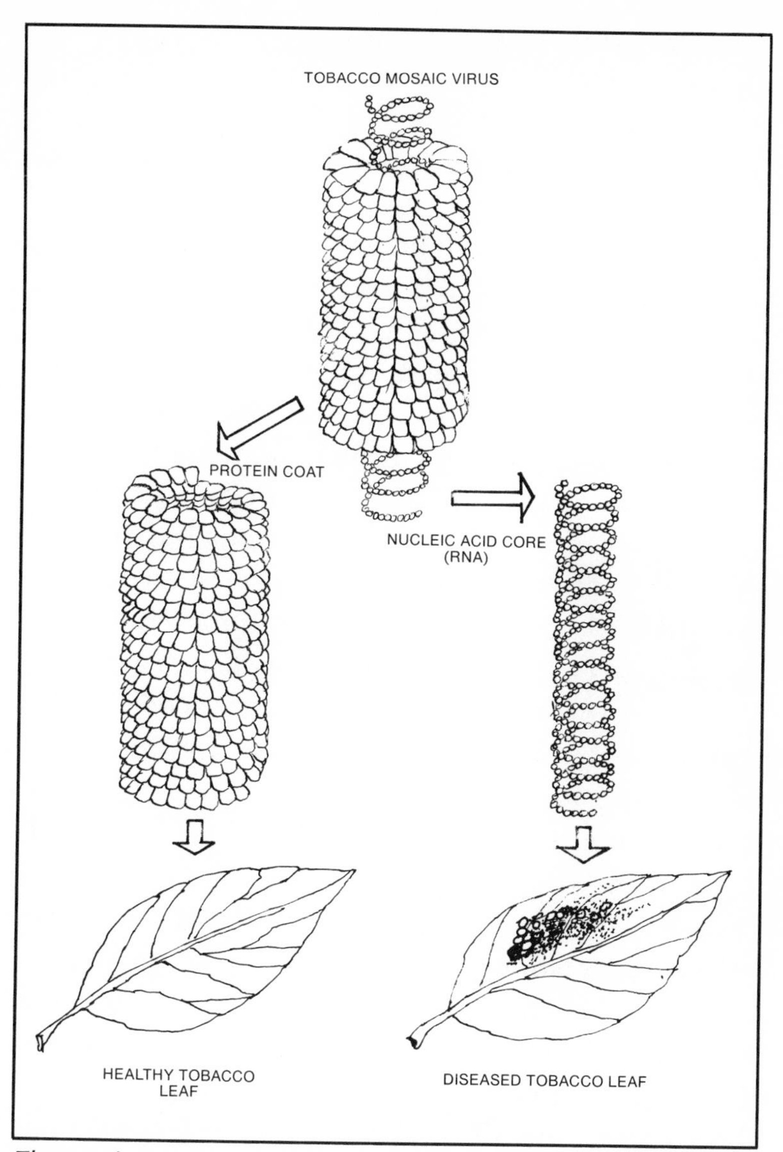

The nucleic acid core (RNA) alone causes the tobacco mosaic virus disease.

shaped and about 3000 Å (one 100-thousandth of an inch) long. It looks like a piece of thick-walled tubing housing a core of coiled wire. The cylindrical wall, which is made of protein, surrounds a delicate coiled fiber of nucleic acid.

Fraenkel-Conrat performed a most delicate and difficult feat in biochemical surgery. He separated the protein coats from the nucleic acid cores. Using a detergent, he dissolved the protein coats from one batch of TMV, leaving a mass of nucleic acid fibers. He then treated another group of TMV particles with a different chemical that ate away the inner nucleic acid core, leaving empty protein containers.

He was now ready for the crucial experiment. Which part of the virus was alive—which part could cause the tobacco disease—the protein coat or the nucleic acid core? A drop of a solution containing the protein coats was rubbed on the leaf of a healthy tobacco plant, and some of the nucleic acid cores on another leaf. After a few days, Fraenkel-Conrat examined the leaves and found all of them spotless. Neither the coats nor the cores, when separated, showed any signs of life. The coats were then mixed with the threads and applied to the leaf. This time yellow spots showed up, identical with those produced by the natural living virus. This meant that the virus was alive—it caused the tobacco mosaic disease. An examination of some of this mixture under the electron microscope revealed coats containing cores. The nucleic acid threads prepared from one batch of viruses had slipped into the empty protein coats of another and were completely at home. These viruses not only looked but behaved as if they had never been taken apart. Evidently both parts of the virus—the coat and the core—were essential for life, or so it seemed at first.

THREADS OF LIFE

Scientific history was being made—a living thing that had been separated into two nonliving molecules was put together again and was alive. Life had been created in a test tube from lifeless chemicals—or so it seemed. Fraenkel-Conrat was very cautious. He had to be certain of the results and so he repeated the experiment again and again and again. More protein coats and nucleic acid cores were prepared and applied to different tobacco plant leaves. This time he varied the amount of material used. No matter what quantity of the coats without cores he added, the tobacco leaves showed no yellow spots; the proteins were lifeless. However, when the amount of nucleic acid cores normally present in the virus was increased a thousandfold and then rubbed into a leaf, the plant developed the tobacco disease. Why did this happen?

Nucleic acid fibers are extremely fragile and are easily broken in the process of preparing them and using them to infect the plant. This is what apparently happened in the earlier experiment. To be infectious, the entire unbroken thread must enter the plant cell. Broken nucleic acid fibers lose the power to cause the disease. When enough fibers are extracted, a few escape injury, remain intact, and are infectious. The coiled core of the virus alone, the nucleic acid molecule, is the thread of life.

Thus the study of the virus led to the basic molecule of life DNA. In the case of the tobacco mosaic virus, the nucleic acid is actually RNA, ribonucleic acid, but it does for this virus what DNA does for most other living things—gives them life. The lowest common denominators of life are the nucleic acids—RNA for a few viruses and DNA for practically all the other organisms.

DNA is the "spark of life," the master molecule that designs all the other molecules of life.

THE CELL—THE HOUSE OF DNA

There are more than a million kinds of plants and animals, and they display tremendous differences in size, shape, and way of living. Nevertheless, all organisms—from amoebas to zinnias—are composed of very similar units, cells. These are the building blocks and the centers of life.

A most remarkable property of a living cell is its ability to maintain its structure and chemical composition in spite of the constant coming and going of billions of assorted kinds of molecules every moment. The tremendous task of preserving the structure and function of a living cell is largely the work of DNA.

With few exceptions, cells are so small they can be seen only through a microscope. Some organisms such as bacteria and amoebas are microscopic in size and consist of one cell. The one-celled organism eats, breathes, digests, excretes, moves, and reproduces. Most organisms consist of a great many cells and the work of living is divided among them. Some cells in your body enable you to move, others carry messages, and still others distribute oxygen. Groups of cooperating cells form tissues, and tissues make up the organs of your body—the heart, liver, kidneys, brain. All of these are members of an organization of 10 trillion cells engaged in keeping you alive.

CELL PARTS

Cells are remarkably uniform in basic design. As seen through an ordinary microscope, most of them have a single central body, the nucleus (NEW-klee-us) embedded in a jellylike layer called the cytoplasm (SIGH-toe-plazm). There is a very thin single membrane around the cytoplasm, the cell membrane, and a double membrane around the nucleus, the nuclear membrane.

CYTOPLASM—THE SEA OF MOLECULES

The cytoplasm is a colorless liquid consisting mostly of water in which there are myriads of molecules—carbohydrates, fats, proteins, minerals, vitamins, and many others. This sea of assorted molecules that bathes the nucleus is the source of all the materials required by DNA to conduct its business. The cytoplasm contains many tiny structures, each with a particular job. For example, there are from 50 to 50,000 bean-shaped structures scattered throughout the cytoplasm that generate the energy needed for doing the work of the cell. These "power plants," called mitochondria (my-toe-KON-dree-uh), are capable of extracting energy from the food molecules floating around in the cytoplasm.

NUCLEUS—DIRECTOR

The control center of the cell is the nucleus. This single spherical or egg-shaped structure contains at least one large granule called a nucleolus (new-KLEE-oh-lis) or little nucleus, and many finer granules referred to as chromatin (CROW-muh-tin), which means color. These granules stain very readily with certain dyes, making them stand out very clearly under the microscope. When the cell is ready to reproduce, the chromatin granules collect into pairs of threadlike structures known as chromosomes (CROW-muh-zomes) or colored bodies.

CHROMOSOMES

The cells of each kind of organism contain a specific and characteristic number of chromosomes referred to as the species number. For humans, the species number is 46 (23 pairs of) chromosomes; for the house fly, it is 12 (6 pairs); and for the corn plant, 20 (10 pairs).

The chromosomes contain the hereditary material of an organism, that is, the material passed on from parent to offspring and responsible for the great resemblance between them. At the turn of the century the chromosome was pictured as containing thousands of hereditary units, the genes, strung together like beads on a necklace. A gene has a definite position on the chromosome and determines the heredity of some specific trait—eye color,

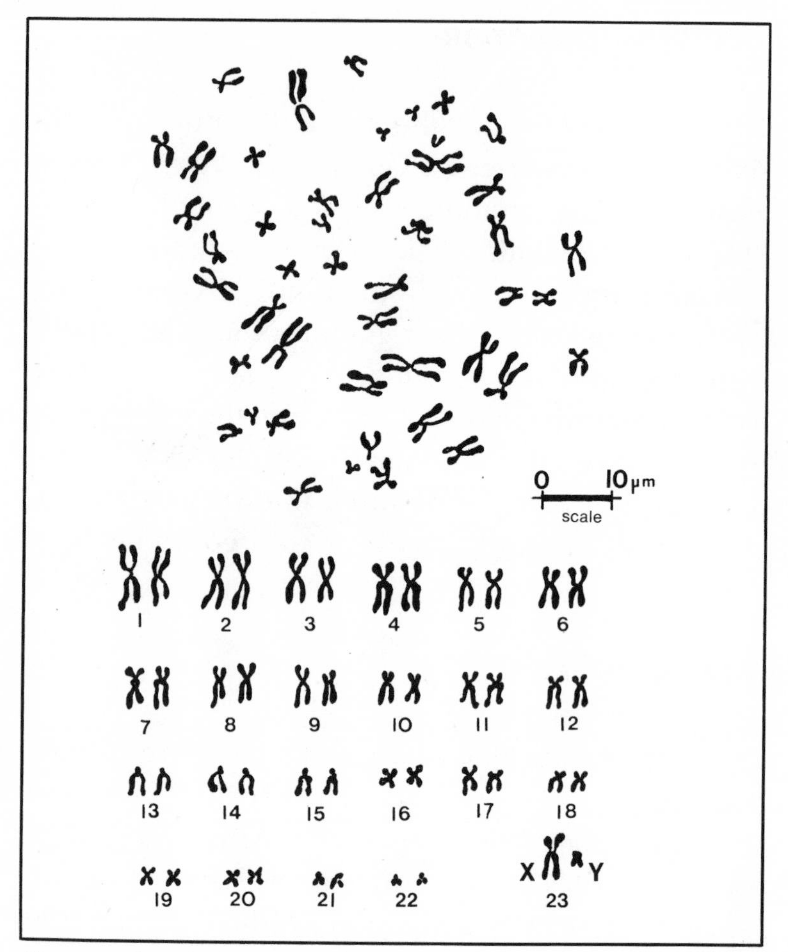

Human chromosomes (male)—23 pairs.

skin color, or blood type. Today scientists visualize the chromosome as a package holding tightly coiled molecules of DNA. The genes are regarded as spots or areas along this long chain of DNA. It is estimated that if the DNA crammed into a single human cell were unwound it would be as long as you are tall and would contain millions of genes.

THE LIGHT MICROSCOPE

Most of what we know about cells and microbes comes from observations made with a light microscope. Since the discovery of the world of microbes in the 17th century by Anton van Leeuwenhoek with his 200-power single-lens magnifier, microscopes have been made more and more powerful. There is, however, a limit to the power of microscopes which use ordinary light and glass lenses. The greatest magnification possible with such microscopes is about 2000 times. At this magnification, the average human cell, which is about 10 micrometers (μm) or one two-thousandth of an inch across, appears to be about an inch long. As important as magnification is the resolving power of a microscope, that is, the ability to distinguish clearly between two objects very close together. The resolving power of the modern light microscope is about 100,000 times; that is, it can clearly distinguish between two lines 0.17 μm 100-thousandth of an inch apart. It enables you to see tiny bacteria but not viruses or the giant molecules of life such as DNA or RNA.

THE ELECTRON MICROSCOPE

The modern electron microscope can magnify a million times and can resolve objects about 2Å or one 100-millionth of an inch apart. By using a beam of electrons, which have a much shorter wave length than that of light, greater magnifications and resolutions can be achieved. The bending and focusing of the electron beam is done by electromagnets instead of glass lenses. Electrons are emitted by a heated tungsten filament similar to that in a light bulb. The electrons stream through a vacuum tube about 4 feet long and pass through the object in the same way that X rays penetrate flesh and bones.

The parts of the object that are dense slow up or deflect the electrons while the less dense parts permit the electrons to pass through. These differences in the density of the various parts of an object and therefore exposure to electrons produce the image. The image may be enlarged and focused

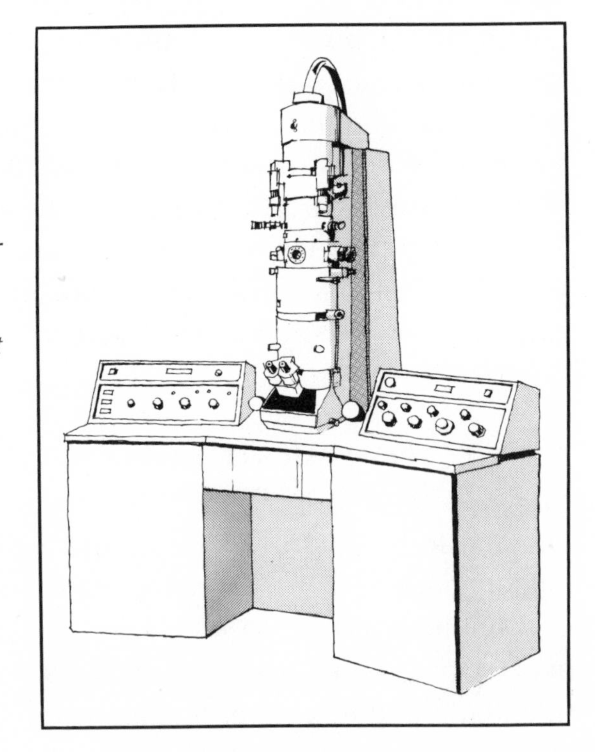

The modern electron microscope can magnify about one million times and can distinguish between objects 2Å apart, which is about half-a-million times greater than that of the human eye.

either on photographic film to make a picture or on a viewing screen similar to that in a television set, where it can be seen directly.

Genes in action can be seen with the electron microscope. DNA molecules have been observed "making" RNA molecules and RNA molecules have been "caught in the act" of producing proteins. Improvements in the electron microscope continue to be made. A new type is the scanning electron microscope, through which you can see minute structures in three dimensions. In this microscope beams of electrons scan the surface of an object and bounce off it instead of passing through. 3-D views of entire organisms and of cell structures are adding to our understanding of these objects.

In 1976 Albert V. Crews at the University of Chicago took motion pictures of single uranium atoms with a special scanning electron microscope he developed. He plans to further improve this instrument for viewing smaller atoms. This achievement points to another possible technique for studying the atomic structure of DNA, RNA, and proteins.

THE CELL THROUGH THE ELECTRON MICROSCOPE

When a cell is viewed through an electron microscope many new and unexpected cellular details are revealed. The cytoplasm, which under the light microscope seems to be a shapeless, formless liquid, at electron microscope magnification appears to be riddled with a network of interconnecting membrane-covered cavities which resemble canals. They are called the endoplasmic reticulum (en-doe-PLAZ-mick re-TICK-you-lum). This system of canals seems to start at the outer cell membrane, meanders through the cytoplasm, and eventually connects with the membrane surrounding the nucleus. It may be the roadway for the transportation of molecules from the outside throughout the cell down to the nucleus. The walls of these canals are covered with thousands of very fine particles called ribosomes (RYE-bo-zomes), so named because they are rich in RNA, ribonucleic acid. Ribosomes are the protein-producing centers of the cell. At electron microscope magnifications, mitochondrions, which are the power plants of the cell, have a covering consist-

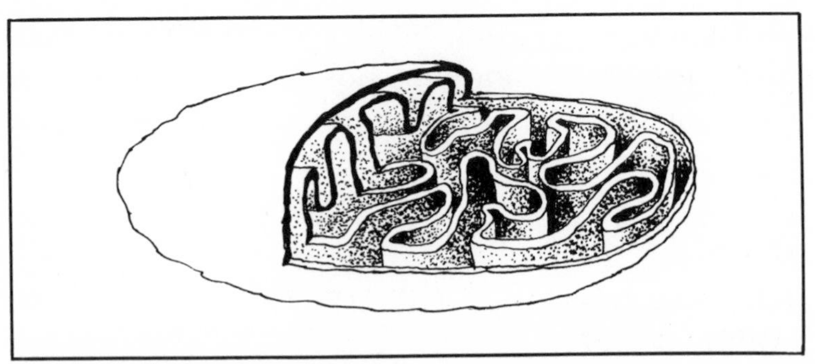

Mitochrondria are the power plants of the cell; these spherical or sausage-shaped structures are found scattered throughout the cytoplasm. The inner walls are folded into a maze of partitions where energy-releasing chemical reactions take place.

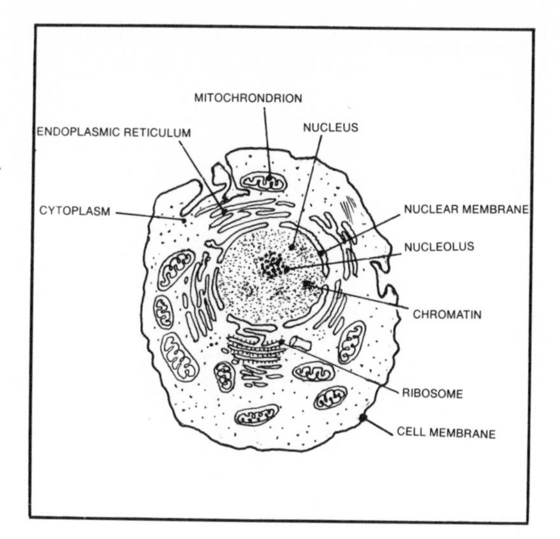

The cell.

ing of two layers. The outer layer is smooth but the inner membrane is folded inward to form compartments. The surfaces of both membranes are sprinkled with thousands of tiny particles that carry out the energy-releasing chemical reactions.* Another interesting detail revealed by the electron microscope is the structure of the nuclear membrane. It also appears to be a double membrane with fine perforations in the layer facing the cytoplasm. Bacteria and viruses have simple chromosomes but lack a nuclear membrane.

INCOMPLETE CELLS

The virus is, in a sense, an incomplete cell and therefore an incomplete organism. It lacks a cell

*There is some speculation based on similarities in structure that mitochondrions were bacteria which invaded the cell a long time ago and have been living there ever since.

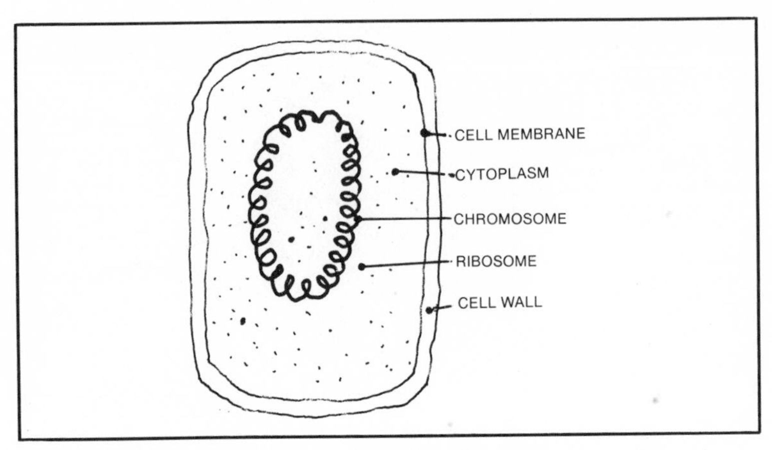

In bacteria, the DNA thread is the chromosome; there is no distinct nucleus with a membrane around it.

membrane and cytoplasm. In chemical composition and in behavior, viruses and chromosomes are much alike. Both are giant molecules of nucleoproteins either of DNA or RNA, and both reproduce themselves only inside cells. The chromosome uses the cytoplasm of the cell in which it is housed as a source of molecules; the virus lives on the borrowed cytoplasm of its host cell. The virus has been aptly described as a hungry piece of chromosome looking for a cell.

CELLS IN TECHNICOLOR

Many structures in a living cell are either invisible or very difficult to see even when they are greatly magnified. With the addition of certain dyes, hidden cellular structures become visible and distinct parts come into sharp focus under the microscope. One such dye is methylene blue, which stains nuclear parts dark blue, particularly the chromosomes, and makes them stand out very clearly. A combination of methylene blue and another dye, eosin (EE-oh-sin), simultaneously colors the nucleus blue and the cytoplasm pink. Dyes are also useful in identifying the chemical composition of cell parts. Acid fuchsin (FEWK-sin) is a dye which colors DNA reddish-purple. When it is applied to a cell, only the chromosomes stain that color, indicating that DNA is located only in the nucleus. Another dye, azure B, colors DNA blue-green and RNA purple. With this dye, the chromosomes stain blue-green and the nucleolus and cytoplasm appear purple. These staining reactions place DNA in the chromosomes and RNA in the nucleolus and cytoplasm.

THE CELL AS A CHEMICAL FACTORY

The cell is a very complex and busy chemical factory carrying on the big business of living. It is a marvel in miniature construction with many built-in parts, each playing its role in the countless activities going on in the cells. We will see in the chapters that follow how the DNA in the nucleus acts as the master planner and master builder, and RNA as the contractor. The ribosomes are the protein-producing machines, the mitochondria the power stations, the endoplasmic reticulum the channels for

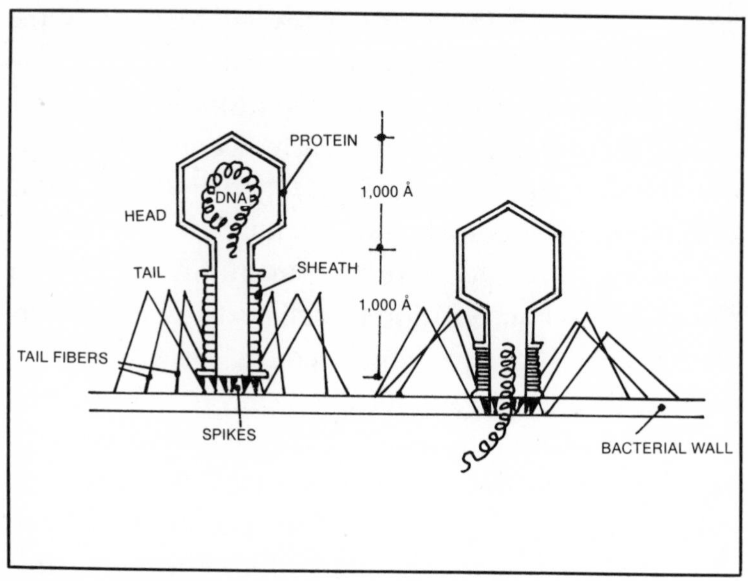

A virus—T2 phage.

transportation of materials, and the cytoplasm the source of myriads of molecules, while the cell membrane is the gateway which controls the movement of molecules in and out of the premises. The basic chemical substances and the chemical reactions of life are our next concern.

THE CHEMISTRY OF LIFE

The drama of life, which is produced and directed by DNA with the able assistance of RNA, takes place in the cells of an organism. The actors are atoms and molecules, with the leading role played by protein molecules. The dialogue is the thousands of chemical processes taking place as the molecules and atoms act and react. To understand this drama and follow the story, it is necessary to be familiar with the language of chemistry.

ELEMENTS AND ATOMS

The letters in the alphabet of basic chemistry are the elements, the basic materials of which everything in the universe is made. An atom is the smallest part of an element. It cannot be broken down into anything simpler and still remain the same substance. There are 105 elements known today, 92 of which are found naturally. The remainder have been produced in the past three decades. Some elements you know by name—oxygen, hydrogen, carbon, nitrogen, iron, mercury, uranium. Others, with names ranging from actinium to zirconium, may be unfamiliar to you. About a dozen elements are very common, and compose most of the things around you.

The smallest particle of an element, iron for example, is an atom of iron. Since there are 105 elements, there are 105 different kinds of atoms. An atom is very small. A quarter of a billion hydrogen atoms, when lined up side by side, would occupy only one inch; each atom is one A in diameter.

COMPOUNDS AND MOLECULES

Atoms tend to join with one another, forming groups referred to as molecules. The atoms in a molecule may be the same or they may be different. A molecule of water, for example, contains three atoms: two of hydrogen and one of oxygen. The atoms in a molecule of each element are always the same. Oxygen, the gas found in air, consists of molecules each of which contains two atoms of oxygen. Molecules in which the atoms are different are called compounds. Water is a compound since a molecule of it contains two different kinds of atoms, namely, two hydrogen atoms and one oxygen atom.

If you think of the 105 elements as the letters in the chemical alphabet, and compounds as the words formed by these letters, it becomes clear that the number of possible chemical words or compounds exceeds that in all spoken and written languages in the world.

CHEMICAL SYMBOLS

The chemist uses a shorthand system for writing the names of the elements and compounds. In this system, the first letter of the name of the element is frequently used as a chemical symbol. For example, O represents oxygen, N stands for nitrogen, C is the symbol for carbon, and so on. Since there are only twenty-six letters in the alphabet and a different symbol is assigned to each element, some are represented by two letters, such as Al for aluminum, Ca for calcium, and Cl for chlorine. The symbols for molecules are derived from the symbols of the atoms they contain, just as words are combinations of letters. A molecule of oxygen with its two atoms is written as O_2, the compound water as H_2O, and carbon dioxide as CO_2. Two molecules of water are $2H_2O$ and four of carbon dioxide, $4CO_2$.

THE ELEMENTS OF LIFE

Living things are made principally of six kinds of atoms. Your body, for example, is 65% oxygen (O) by weight, 18% carbon (C), 10% hydrogen (H), 3% nitrogen (N), 2% sulfur (S), and 1% phosphorus (P). These six elements account for 97% of your weight. About twenty-five additional elements make up the rest of you. They include calcium, potassium, sodium, chlorine, magnesium, iron, iodine, fluorine, copper, zinc, and cobalt.

AIR	LAND	SEA	LIFE
OXYGEN	OXYGEN	OXYGEN	OXYGEN
NITROGEN			NITROGEN
	SILICON		
	ALUMINUM		
	IRON		
	CALCIUM		
		HYDROGEN	HYDROGEN
		CHLORINE	
		SODIUM	
			CARBON
			SULFUR
			PHOSPHORUS

The chemical elements around us.

The six elements of life—C, O, H, N, S, and P—are neither rare nor restricted to living things. Air is almost entirely O_2 and N_2, water is H_2O, and the other elements are in the earth—C in coal and S and P in rocks and soil.

A distinctive feature of living material is the kinds of molecules they form from these elements. The molecules associated with life are usually large, complicated, and fragile. They contain hundreds, thousands, and sometimes millions of atoms. One of the largest known molecules, the tobacco mosaic virus, is a giant molecule of nucleoprotein containing over 5 million atoms. There are, however, only six kinds of atoms in this mammoth molecule—C, O, H, N, S, and P. The molecules of the nucleic acids, DNA and RNA, may contain hundreds of thousands of atoms but they are limited to only five kinds—C, O, H, N, and P.

ATOMIC STRUCTURE

The kinds of atoms that combine and the way in which they join are governed by chemical rules and regulations. To understand these chemical reactions, you must know something about the structure of atoms.

All atoms, with the exception of hydrogen atoms, consist of three basic particles—electrons, protons, and neutrons. Electrons are negatively charged particles that whirl around a dense central core or nucleus containing the protons and neutrons. Protons are positively charged particles. Neutrons have no charge.

Each kind of atom has a different number of basic particles. Hydrogen, for example, consists of just one electron and one proton; carbon contains six electrons, six protons, and six neutrons; nitrogen has seven of each, and oxygen eight. Since the number of electrons in an atom is equal to the number of protons, the atom is electrically neutral.

ELECTRONS IN ORBIT

The electrons revolve around the nucleus of an atom along definite paths or orbits located at varying distances from the nucleus. Each orbit can hold only a certain number of electrons. The first orbit, which is nearest the nucleus, has no more than 2 electrons,

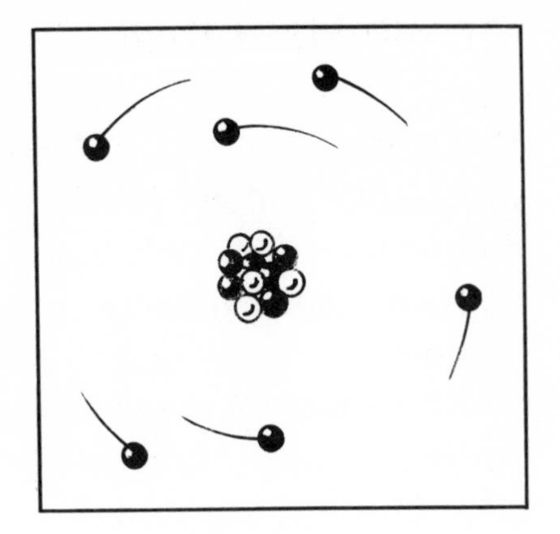

A carbon atom contains six electrons revolving around a nucleus consisting of six protons and six neutrons.

the second orbit, which is further out, is limited to 8, and the third holds as many as 18.

The single electron of H occupies the first orbit. The six electrons of carbon are distributed so that two are in the first orbit and the other four in the second. Hydrogen lacks one to complete its outermost ring. Carbon lacks four, and oxygen, with six electrons in its outermost orbit, lacks two. The behavior of a particular kind of atom in relation to other atoms is influenced by the number of electrons in its orbital rings. Atoms tend to fill up their outermost orbits by either taking electrons from other atoms, or giving up all the electrons in their outer ring to another atom, or sharing their electrons with another atom. In general, elements with less than four orbital electrons in the outer ring (if the maximum is eight) tend to lose them to other atoms, those with more than four tend to take electrons from other atoms, and those with four tend to compromise, being neither lenders nor borrowers but sharers of electrons. Atoms capable of exchanging or sharing electrons can undergo chemical reactions and form new molecules. The energy holding the atoms together in the molecule is called a chemical bond. Each atom has a definite and limited number of chemical bonds, which may be thought of as links connecting it with other atoms.

CHEMICAL BONDS

Since we are primarily interested in living organisms and the chemical reaction of life, let us examine the chemical bonds of the six elements of life. The hydrogen atom has only one chemical bond and it can be represented as H–. Since carbon has four outer orbital electrons and room for four more, it has four chemical bonds and is therefore symbolized as

$$-\overset{|}{\underset{|}{C}}- .$$

Oxygen, with six electrons in its outermost ring, has room for two more and it therefore is represented with two bonds as –O–. Nitrogen, with five electrons, has three chemical bonds and is represented as

$$-\overset{|}{N}- .$$

Similarly, sulfur with two chemical bonds is –S– and phosphorus with three is

|
–P– .

MOLECULES—BONDED ATOMS

It is now possible to understand how some very simple molecules are formed. For example, H– can combine with another H– to form H_2, which may also be written as H–H. H_2 is the molecular formula which tells you that a molecule of hydrogen contains two atoms of hydrogen. H–H is the structural formula which gives you some idea of the arrangement of the atoms in the molecule. Four H–'s can also join with

to form CH_4, which is known as methane, a gas found in swamps and in natural gas used as fuel.

|
–N–

can hook onto three H–'s to form

H
|
H–N–H

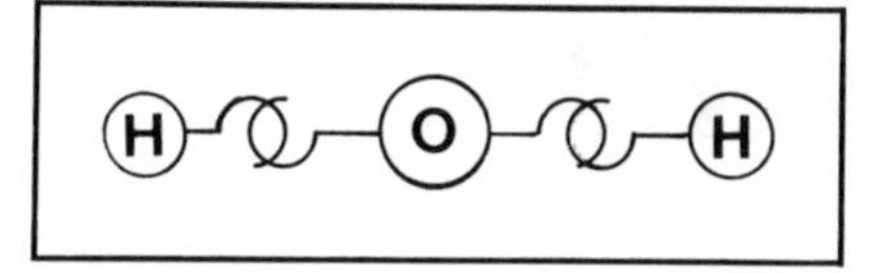

A water molecule, H_2O, contains one atom of oxygen (O) linked to two atoms of hydrogen (H).

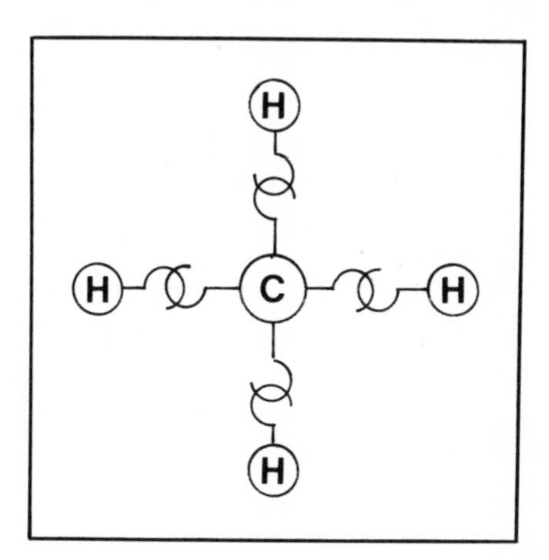

A methane molecule contains four hydrogen atoms joined to one carbon atom—CH_4.

or NH_3, ammonia, a gas that is present in household cleaners.

Atoms may be linked not only by single bonds but also by double or triple bonds. A molecule of oxygen is O=O, nitrogen is N≡N.

CARBON COMPOUNDS

The large complex molecules so characteristic of living matter—such as carbohydrates, fats, proteins, and nucleic acids—are carbon compounds. They are built around a backbone of carbon atoms arranged in a chain. Although carbon has four chemical bonds, which can join with four other atoms, it exhibits the properties of self-linkage, more so than any other atom. Carbon atoms have the tendency to join one another, forming long stable chains or rings.

CARBON CHAINS

Starting with a simple one-carbon compound, methane (CH_4), you can build a two-carbon chain by joining another methane molecule to it. By linking a third methane to the two-carbon chain, you can

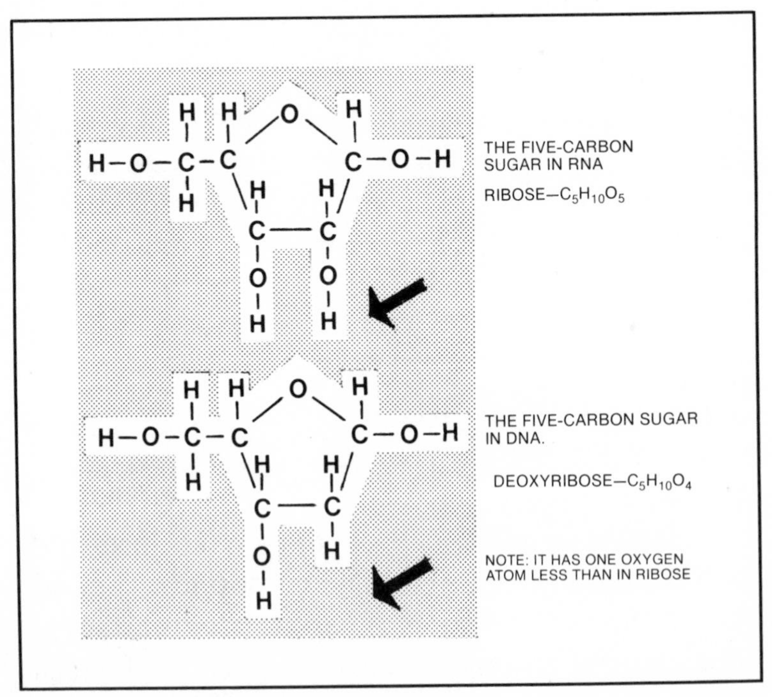

The sugar in the nucleic acid molecules.

extend it to a three-carbon linkage. This process can be extended on and on and the chain may grow in a straight line or it can become branched.

The sugar commonly found in fruits has a six-carbon chain, $C_6H_{12}O_6$. The sugars present in the nucleic acids, ribose and deoxyribose, are both five-carbon sugars—$C_5H_{10}O_5$ and $C_5H_{10}O_4$ respectively.

RINGS OF CARBON

The carbon atoms at one end of the chain may join with those at the other to make a closed ring. The ring may contain six carbon atoms connected by alternating single and double bonds. One of the simplest of these ring compounds has a hydrogen atom attached to each of the six carbon atoms; it is C_6H_6, benzene.

The benzene molecule consists of six atoms of carbon joined in a ring to which six atoms of hydrogen are attached (C_6H_6).

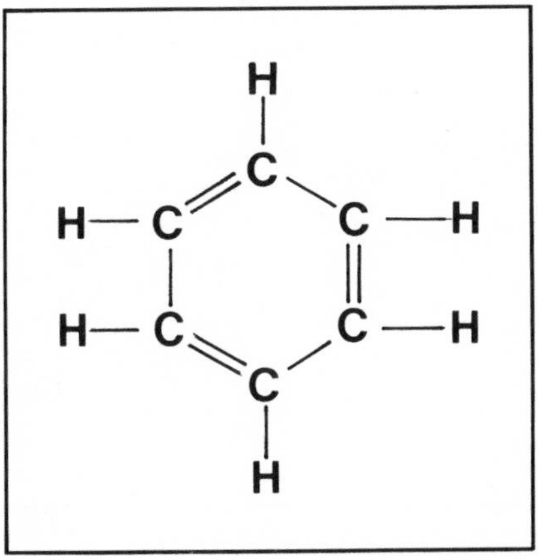

H
C
H— C   C —H
H— C   C —H
C
H

The ring may contain nitrogen or some other elements in addition to carbon. There may be four carbon and two nitrogen atoms in the ring. This is the basic structure of the pyrimidines (pih-RIM-ih-

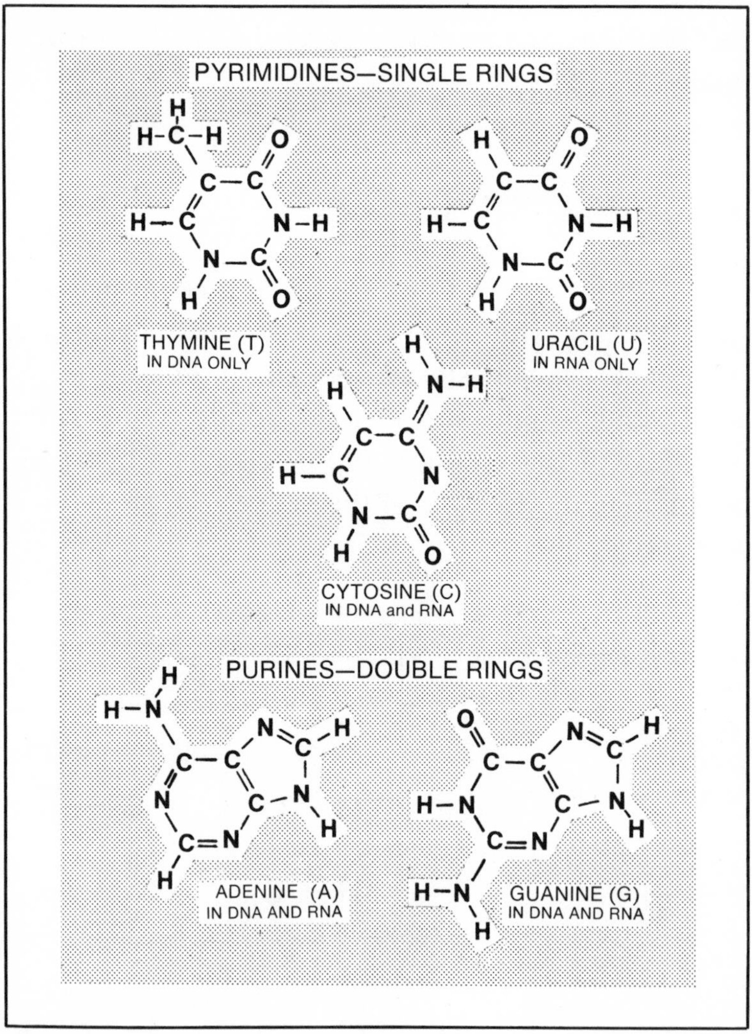

Bases in DNA and RNA.

deans) such as thymine (THIGH-mean), which is part of the DNA molecule, uracil (YOUR-a-sill), found in RNA, and cytosine (SIGH-toe-seen), present in both DNA and RNA.

Another arrangement is a combination of rings such as is found in the purines (PURE-eens). Adenine (ADD-uh-neen) and guanine (GUAH-neen) are such double-ringed molecules found in DNA and RNA.

Now that you have been introduced to some of the basic chemical reactions of life and some of the molecules involved, you can begin to understand the importance of molecular architecture, particularly of DNA, the basic molecule of life. This is our next consideration.

DNA—THE LADDER OF LIFE

The Nobel prize in medicine or physiology for 1962 was awarded to an American and two British scientists for their "discovery of the molecular structure of deoxyribonucleic acid, DNA, which contributed to an understanding of the basic life processes." The winners were James D. Watson of Harvard University, Cambridge, Massachusetts; Francis H. C. Crick of Cambridge University, Cambridge, England; and Maurice H. F. Wilkins of Kings College, London, England.

Their discoveries have been hailed as perhaps the most significant advance in biology in the 20th century. To the world of science these men were no strangers, and this recognition came as no great surprise. The Crick-Watson model of DNA, which looks like a twisted ladder, has been a familiar symbol since 1953 when it was first suggested by this team. Their model has helped solve the mystery of the architecture of this all-powerful, all-purpose molecule and to explain how DNA governs life. It provides us with a physical explanation of how DNA carries hereditary information and also how this information is passed on from cell to cell and from organism to organism.

THE MIGHTY MIDGET

All the DNA in a fertilized human egg cell is found in 46 microscopic chromosomes confined to the nucleus. Nevertheless, the DNA in this single cell holds all the hereditary information necessary to guide its development into a full-grown man or woman. Although DNA is unbelievably small, it contains an incredible amount of genetic information, enough to fill a thousand-volume encyclopedia.

X-RAY PICTURES OF DNA

By 1950 DNA was recognized as the master molecule containing genetic instructions that determined the form and function of each living thing. Scientists became intrigued with the atomic architecture of this molecule, that is, how the atoms and groups of atoms are arranged in DNA. They felt that knowledge of its composition and an understanding of how it is put together would provide the key for explaining how it works.

Electron microscope studies revealed little about its architecture, since DNA is incredibly thin, about 20 A or one ten-millionth of an inch thick, and at electron microscope magnifications it looked like a stiff piece of string without internal structure.

The shape and construction of this molecule, too small to be seen with the most powerful microscope, was explored by another method, which

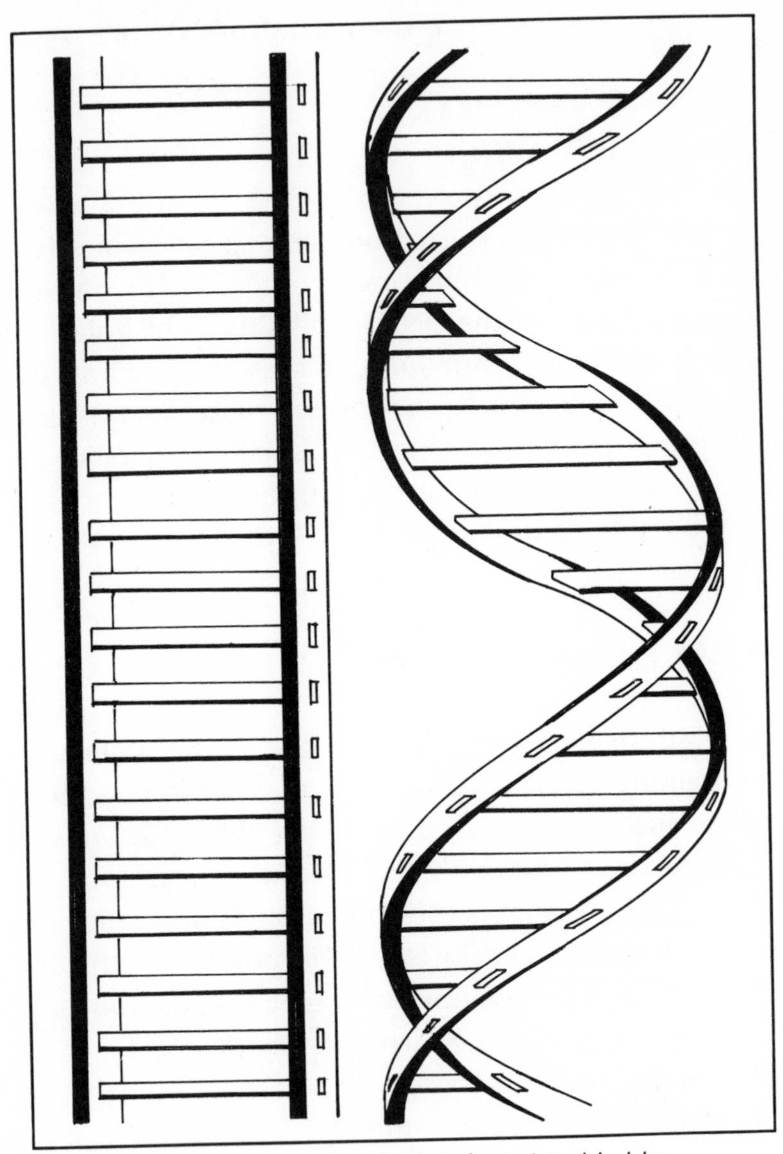

The DNA molecule has the shape of a twisted ladder.

involved studying its shadow. You can place a light source behind an object and cast its shadow on a wall or screen. Suppose you are on the other side of the screen and cannot see the object—your brother, for example. You could probably identify him by moving him around and obtaining shadows from different angles.

This is essentially what Wilkins did; however, he used X-rays instead of light to study samples of pure DNA obtained from various organisms. By bouncing X rays off the atoms in the DNA molecule, Wilkins obtained photographs of their atomic architecture. These are not pictures in the ordinary sense of the word but shadow or silhouette patterns. When X rays strike an atom or group of atoms they are scattered or deflected from their paths and cast shadows on a photographic plate. Groups of regularly repeating or orderly arranged atoms can be identified by the shadow pictures they produce. The X-ray pictures suggested that the DNA molecule is twisted into a spring or coil resembling a spiral staircase or twisted ladder. The steps are flat units stacked one above the other perpendicular to the length of the threadlike molecule, like the steps in a circular staircase. The shape of the DNA molecules of all organisms is the same, regardless of whether they come from a bacterium, fish, or human.

SIX PIECES

Chemical studies revealed that DNA is one of the largest molecules known, composed of many thousands of atoms. However, these huge molecules are constructed from smaller submolecular units, which are used over and over again as building blocks.

DNA contains six kinds of submolecular building blocks. One of them is the sugar deoxyribose, the D in DNA. Another is phosphate, a group of four oxygen atoms and one phosphorus atom. The other four units are purines and pyrimidines, rings of carbon and nitrogen atoms referred to in the previous chapter. They are adenine, guanine, thymine, and cytosine, collectively known as bases. We shall refer to them by the first letters of their names—A, G, T, and C. (Note: In the next few pages we will be using C to mean cytosine rather than carbon.) Sugars, phosphates, and bases—these are the structural parts. But how are they assembled in the DNA molecule?

MOLECULAR MODEL MAKERS

In 1952, Crick and Watson undertook the task of constructing a three-dimensional model of DNA. They thought the best way to determine the structure of this molecule was to build a scale model of it. The only clues they had were the X-ray pictures taken by Wilkins suggesting that it had the shape of a coiled spring or twisted ladder. They also knew that DNA contained only six kinds of submolecules. The most unusual part of their venture was the kind of laboratory equipment they used. It consisted of pieces of metal cut in the size and shape of the six building units of DNA and wire to act as the chemical bonds for joining the pieces.

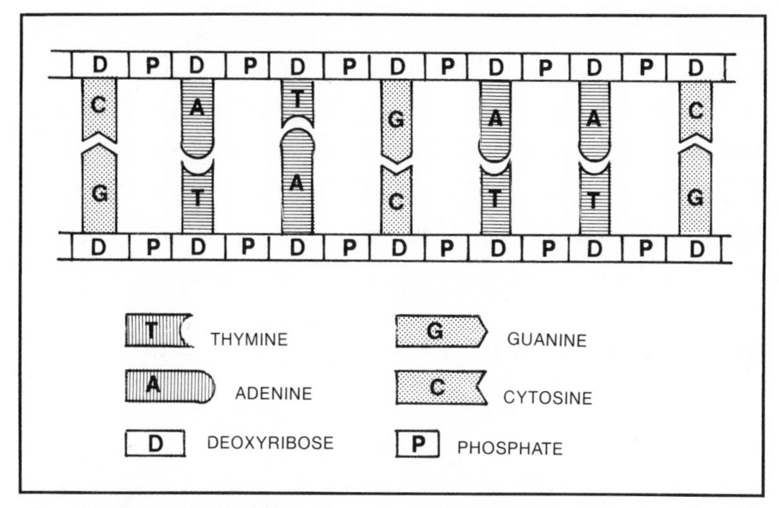

The structure of DNA: How the six building units are assembled in the DNA molecule.

For a year these model makers struggled with this chemical jigsaw puzzle, using imagination and ingenuity to get the pieces to fit together according to all that was known about the chemical and physical structure of DNA. Finally, the pieces fell into place and they came up with their 3-D scale model.

CRICK-WATSON MODEL

How do the six building units fit into the model? The model that Crick and Watson assembled in 1953 resembles a twisted ladder. The sides of this ladder, which are two long threads wound around each other, contain alternating sugar and phosphate units. The rungs, which are flat units stacked one above the other, are attached at right angles to the sugar units in the two twisted threads. Each rung consists of a pair of bases, either an A and a T, or a C and a G. No other combination of bases is found in the rungs of the DNA ladder. The A-T and C-G pairs of bases make rungs of equal length and of the

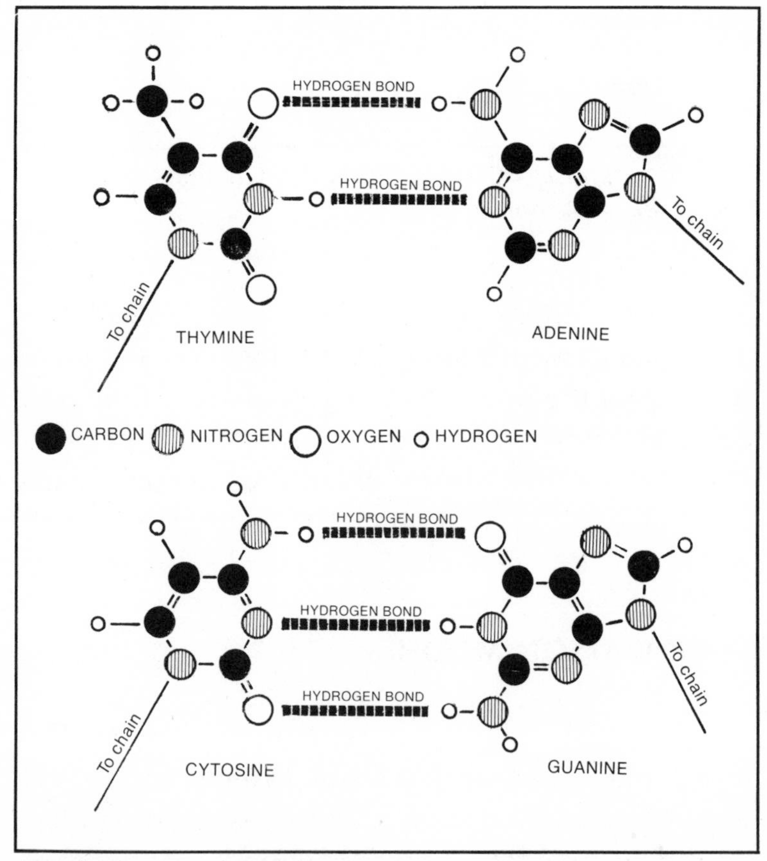

Matched pairs of DNA bases: In the DNA molecule, the base thymine (T) is linked to adenine (A) by hydrogen bonds and the base cytosine (C) is bonded to guanine (G).

proper size to fit between the parallel sides of the ladder; other combinations are either too large or too small. Also, only A can join with T, and G with C, because these bases have matching shapes, like pieces in a jigsaw puzzle. The matched pairs of bases are fastened together by hydrogen bonds, A and T by two, and C and G by three.

Thus an A in strand I, the left side of the ladder, is hydrogen-bonded to T in strand II, the right side of the ladder. T in strand I is paired with A in strand II. G in strand I is wedded to C in strand II, and C in strand I is connected to G in strand II. A is bonded to T, and G to C. Hence the order of the bases in one strand fixes the order of the bases in strand II. For example, when the bases in strand I are CATGAAC, the order of the bases in strand II is GTACTTG.

The ground rules governing the structure of the Crick-Watson model of DNA may be summarized as follows:

1. DNA consists of two intertwining strands which resemble a twisted ladder.
2. Each of the strands consists of a chain of alternating sugar and phosphate units.
3. The individual rungs of the ladder, stacked one above the other and at right angles to the twisted threads, contain pairs of bases.
4. There are only four kinds of rungs: A-T, T-A, C-G, and G-C.
5. The sequence of bases in one strand determines and fixes the sequence of bases in the second.

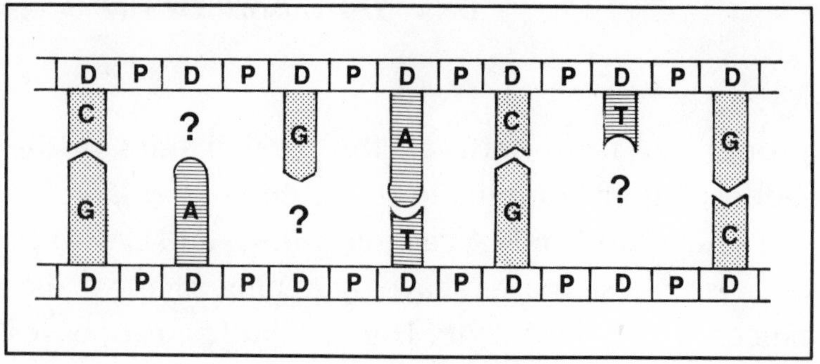

The DNA puzzle: Which bases are missing?

Now that you are familiar with the Crick-Watson model of DNA, let us see how it can be used to explain some of the fundamental problems of life, such as how DNA carries genetic information and how this information is duplicated.

DNA—THE MOTHER MOLECULE

A unique feature of DNA—one that distinguishes it from all other molecules—is the ability to duplicate itself. This "mother" molecule can make another exactly like itself and the duplicate may be passed on to its descendants. These molecules are links in the chain of life that stretches back in time for as long as the species has existed. Each new cell or organism begins life with a set of hereditary instructions built into the structure of its DNA molecule which "tells" it what kind of an organism it is, what to do, and when to do it.

Self-duplication is a fundamental property of life. A bacterium reproduces every 20 minutes by simply splitting in half. Every time it reproduces, its DNA duplicates; the copy containing the blueprints for making another bacterium like itself goes to the new organism. Similarly, each new generation of human beings receives the genetic instructions for human beings from the previous generation.

DNA DUPLICATION IN THE VIRUS

The self-duplicating powers of the DNA molecule can be seen in viruses that feed on bacteria. Such viruses are called bacteriophages (bak-TEAR-ee-ofay-jez), or bacteria eaters, or phages for short. Viruses, you remember, are the smallest and simplest bits of life capable of growing and reproducing, but only inside the cell of another organism. These parasites are giant molecules of nucleoprotein, with a protein coat enclosing a nucleic acid core. The phage called T2 lives inside bacteria which commonly inhabit the human intestines. T2 is tadpole-shaped with a six-sided hollow head and a long hollow tail at the end of which are several fibers. Inside the head there is a single coil of DNA. The phage is about 2000 Å or one 100-thousandth of an inch long. About 1000 phages can fit nicely into a single intestinal bacterium, which is called *Escherichia coli* or *E. coli.*

A single phage attaches itself to a bacterium by means of its tail fibers, and with the help of enzymes it bores a hole through the thick wall of its victim. The tail then contracts and the phage shoots its DNA thread into the bacterial cell in the manner of a hypodermic needle. The head and tail remain outside the cell and only the DNA thread enters. Once inside, the virus DNA takes over the chemical machinery of the bacterial cell like an invading dictator and converts it to its own use. The first step in this "takeover" is to destroy the control center of the cell, the DNA of the bacterium itself. The virus makes enzymes which crumble the bacterial DNA to pieces. The captured cell now works for the DNA of the virus. It continues to take in food, produce energy, and make proteins, but under the directions

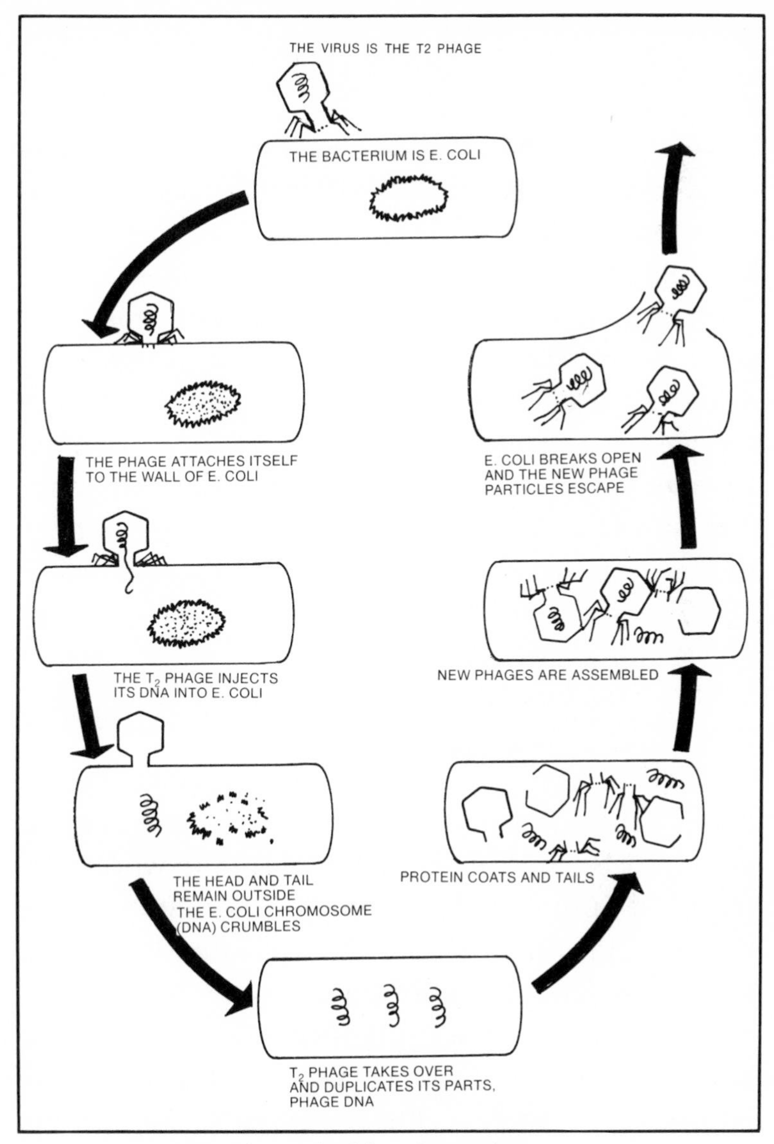

How a virus reproduces inside a bacterium.

and supervision of the conqueror. Virus DNA orders the vanquished cell to make more viruses instead of bacterial parts. Within minutes, several hundred new virus DNA threads are assembled, replicas of the invading DNA. The cell also constructs protein coats, copies of the original virus coat, to fit the virus DNA threads. Ten minutes after the virus first enters the cell, the new DNA threads begin to slip into their new protein outfits. Thirty minutes after the invasion, some 300 new fully clad viruses pack an empty bacterial cell. They then crash through the wall with the help of enzymes, leaving an empty shell behind. Each virus is now ready to inject its deadly DNA coil into another bacterium and duplicate itself until all the bacteria are destroyed.

VIRUS COATS AND COILS

How did scientists prove that invading phages leave their coats outside? A convincing experiment in 1952 by Alfred D. Hershey and Martha Chase, working at the Cold Spring Harbor Genetics Laboratory of the Carnegie Institute, proved that only the DNA of the virus T_2 enters and reproduces inside the bacterial cell. They tagged the protein coat of the virus with radioactive sulfur and the DNA coil with radioactive phosphorus. Since sulfur is present only in the protein and not in the DNA, and phosphorus is limited to the DNA and absent from the protein, the fate of these tagged parts could be followed along the invasion route.

The two batches of tagged viruses were allowed to infect *E. coli* separately for a few minutes. The viruses and bacteria were then separated by an ordinary kitchen blender. Hershey and Chase found that in those few minutes of contact, the viruses had injected "something" into the bacteria, since samples of these bacteria, when allowed to grow, burst and liberated new viruses in the usual way. However, when the sulfur-labeled viruses were used the bacteria were not tagged, that is, the radioactive sulfur remained in the coats of the invading viruses

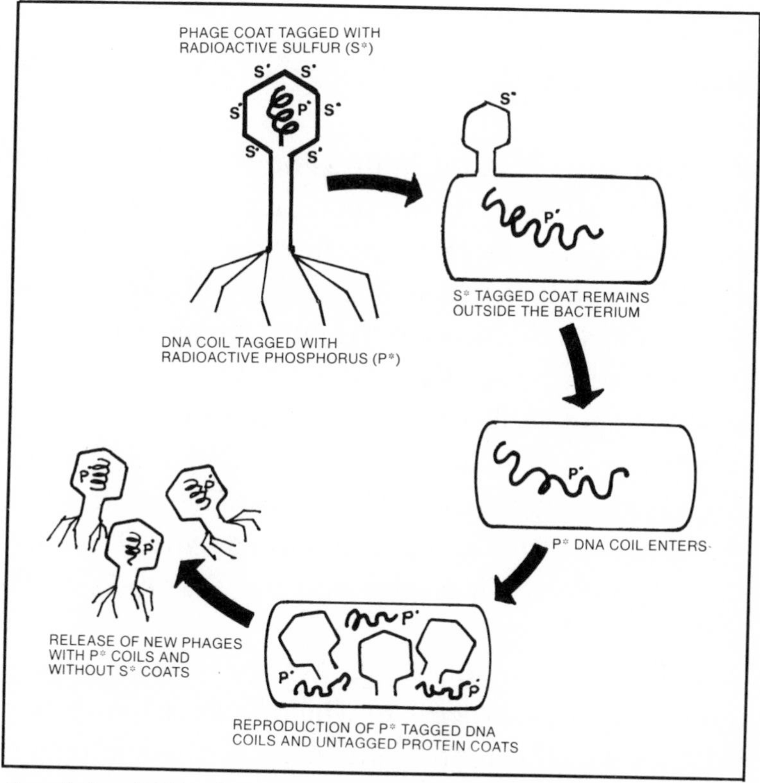

Proof that DNA of a virus is the hereditary material.

and did not enter the bacteria. With the phosphorus-labeled viruses, the bacteria were radioactive, indicating that the phosphorus-labeled DNA had passed into the bacteria. Here was proof that only the DNA enters the bacterial cell and only DNA carries the genetic information of the virus which directs its reproduction inside the bacterium.

For this work, Hershey was one of three to share the 1969 Nobel prize in medicine or physiology.

DNA DUPLICATION—TWO FROM ONE

A most challenging question is how the DNA molecules are able to make such accurate copies of themselves. Let us take another look at the Crick-Watson model, since it suggests a simple mechanism for DNA duplication. As we have already seen, DNA has the shape of a twisted ladder the two sides of which are long chains of alternating sugar and phosphate units held together by connecting rungs. The rungs contain pairs of bases joined by hydrogen bonds; A in one chain is coupled with T in the second chain or vice versa and the C with G, or G with C.

Crick and Watson suggested that DNA duplication occurs by a process in which the twisted ladder splits down the middle and each half assembles a complete ladder. First the DNA ladder begins to untwist and the hydrogen bonds holding the base pairs together break in orderly succession starting at one end. In effect, the ladder straightens out and unzips lengthwise through the middle of each rung, thus separating the bases. The members of each pair of bases are now exposed to the cytoplasm, the cell's sea, which contains the molecular building blocks for DNA as well as other raw materials of life. Each base in the half ladder is now free to pick up a new molecular mate from the cytoplasm. However, these chemical matings are limited. For example, A in the chain can join up only with a free T, since these two bases fit together perfectly. Free C and G bases will be rejected by A because they do not match. Similarly, G in the chain hooks up with a free C base or the reverse. The newly attached bases pick up the sugar and phosphate units to complete the ladder. In effect, each base latches on to a new partner exactly like its previous one. The new units are in precisely the same place and in the same order as the old ones.

Finally, when the splitting and the pairing processes are completed, two DNA molecules stand in the place of one. Each of the two halves in the original DNA molecule has assembled a new half which is exactly like the old one. Instead of one ladder, there are two ladders, each containing one original and one new chain. The resulting DNA molecules are identical with the original mother molecule. When the cell reproduces, each daughter cell receives DNA molecules in the same amount and of the same kind as those in the original cell.

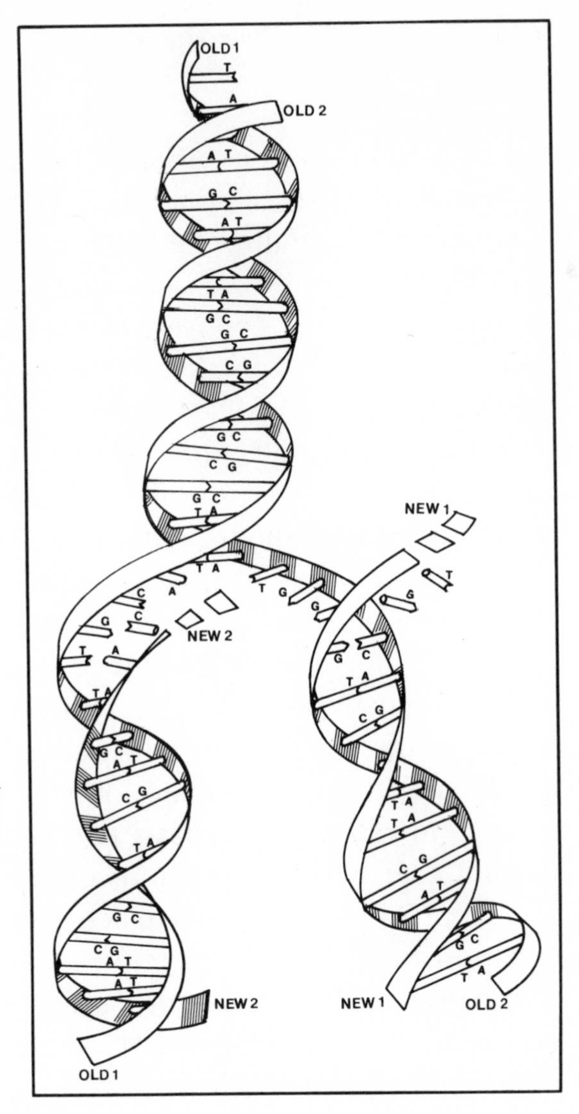

DNA duplication—two from one. The DNA ladder untwists and unzips through the hydrogen bonds linking the pairs of bases. The bases in each half of the ladder pick up new molecular mates from the cytoplasm and assemble a new partner exactly like its previous mate. Two ladders are formed, each consisting of an old and a new strand.

ONE OLD, ONE NEW

How do we know that DNA ladders make other ladders like themselves by first splitting and then each half building a full one?

A sophisticated experiment to answer this question was devised by Matthew S. Meselson and

Franklin W. Stahl at California Institute of Technology in 1958, using bacteria grown on food enriched with heavy nitrogen N^{15}. They succeeded in raising a population of bacteria in which both strands of the DNA molecules in them were tagged with N^{15}. These tagged bacteria were then put on a diet containing ordinary nitrogen N^{14}, and permitted to reproduce for two successive generations. After the first division, the DNA content of these organisms was analyzed and found to be half N^{15} and half N^{14}. After the second division, two kinds of DNA appeared; one was all N^{14} and the other half N^{15} and half N^{14}.

These findings are in complete accord with the Crick-Watson explanation of DNA duplication. At the first division, the two N^{15} strands of DNA separated and each became the mold for assembling a new strand from the N^{14} food around it. Each daughter cell got a DNA molecule consisting of one N^{15} and one N^{14} strand. Just before the next cell division, the $N^{14}:N^{15}$ strands separated and each assembled another N^{14} strand. When this cell divided, one of the new cells received the $N^{15}:N^{14}$ molecule and the other cell got the $N^{14}:N^{14}$ molecule.

How does this experiment show that DNA ladders split in half and each half builds a new one? When the tagged ladders split, this resulted in two tagged half ladders. These then built untagged halves, so that each of the ladders was half tagged and half untagged. These half-and-half-type ladders then split and again each half built another untagged

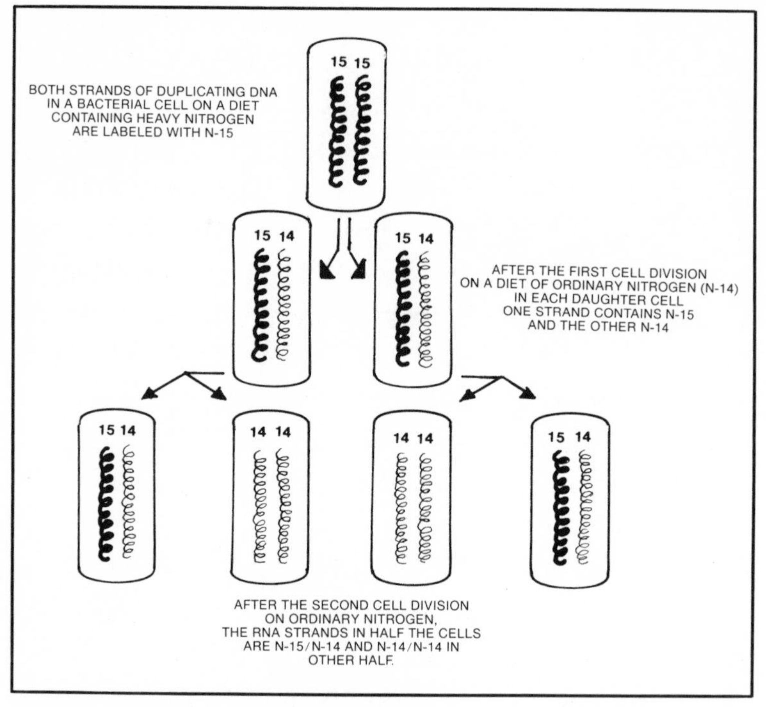

DNA duplication and cell division.

half. This time two kinds of ladder appeared. One type was half tagged and half untagged; in the other kind both halves were untagged.

DUPLICATING TAGGED CHROMOSOMES

Before a cell divides, each chromosome duplicates itself and appears as a pair of short rods lying side by side. When the cell divides, the pairs separate and one chromosome of each pair goes to each of the daughter cells. Thus, each new cell contains copies of each of the original chromosomes. A question that arises is "How is the reproduction of the chromosomes related to the reproduction of the DNA in it?" Another experiment similar to that with bacteria tested the template theory of DNA duplication in terms of chromosome duplication.

In 1957, J. Herbert Taylor of Columbia University in New York City, working with Walter L. Hughes and Philip S. Woods of Brookhaven National Laboratory, attempted to find an answer to this question by following the fate of tagged DNA through successive divisions of the chromosomes in the cells of a bean plant. When cells are grown in a solution containing thymidine, which is one of the constituents of DNA, the thymidine is taken up only by the chromosomes and the DNA in them, since no other part of the cell uses it. If this molecule is now tagged with tritium, a radioactive form of hydrogen, it can be traced into the chromosomes and through their successive divisions. The rays emitted by tritium travel a very short distance, and when they strike a photographic plate dark spots are produced, pinpointing the exact location of DNA in the chromosome. Although a tagged chromosome cannot be distinguished from an untagged one by the human eye, photographic film can "tell" them apart. This technique is called radioautography (radio-aw-TOG-ra-fee), because molecules tagged with radioactive atoms take pictures of themselves on photographic film.

The roots of bean plants were grown in a solution of tagged thymidine for a few hours. Taylor then removed a few of the bean cells, squashed them under a plate of glass, and covered it with a thin sheet of photographic film. The radioactive rays produced a photograph showing where the thymidine was concentrated. The radioautographs (radio-AW-toe-graphs) or pictures showed that all the chromosomes in the cell were tagged and the radioactive molecules were equally distributed. The plants were now taken out of the radioactive solution and placed in a solution of nonradioactive thymidine. After these labeled chromosomes had

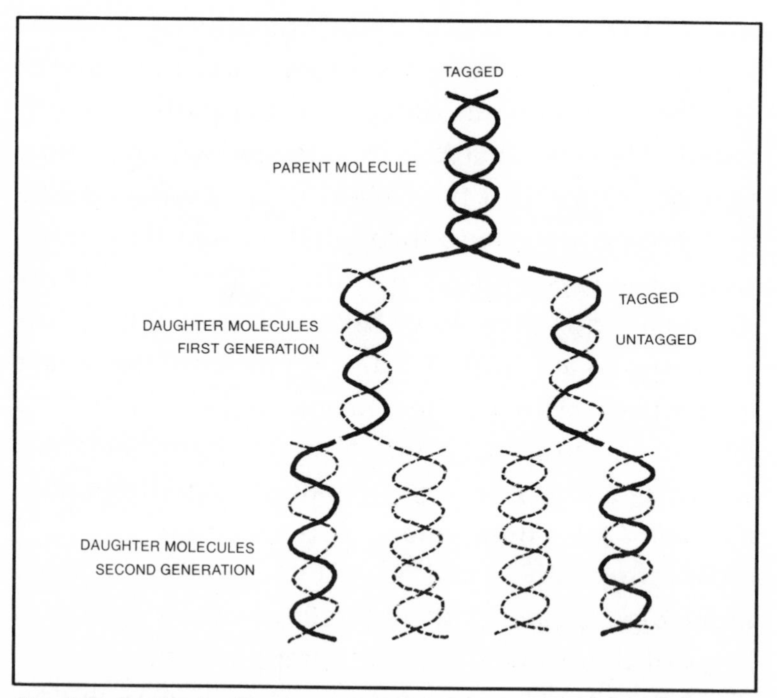

Tracing DNA duplication: DNA ladders split and each half builds a new ladder.

reproduced, some of the cells were again squashed under glass and covered by film. This time photographs showed that in each of the paired chromosomes one was tagged and one was not.

Taylor explained this meant that a chromosome consists of two parts or strands, each of which acts as a mold or template. When the cells were in the radioactive solution, each pair of chromosomes, after splitting in two, assembled a new radioactive mate. All the new chromosomes contained one tagged and one untagged strand. When transferred to the nonradioactive solution, each of these strands formed untagged mates. Hence half the chromosomes were tagged and half were not. This is exactly what happened in the experiment with bacteria.

These findings are consistent with what is known about the DNA molecule and the Crick-Watson theory. The DNA molecule contains two chains twisted around one another. As the chains unwind each builds itself a new partner. Duplication goes on as the strands separate and each serves as a template for assembling a new partner. We still do not know how DNA molecules are arranged in the chromosomes.

TEST-TUBE "DNA"

One of the most exciting experiments supporting the ladder-splitting, ladder-building theory of DNA duplication came from efforts to make DNA in a test tube. In 1956 Arthur Kornberg directed a team of researchers at Washington University, St. Louis, Missouri, in the successful synthesis of a DNA-like molecule in a test tube outside the cell. This feat was accomplished with the help of an enzyme isolated from rapidly growing *E. coli* by the Kornberg team. This DNA duplicating enzyme was named DNA polymerase (POE-limb-er-ace) or the Kornberg enzyme. In the first attempt, the four DNA bases (A, C, T, and G) were mixed with sugar, phosphate, and some other chemical including DNA polymerase, the enzyme for putting the pieces together. The results were disappointing—only a tiny bit of DNA-like substance was formed—but that was enough to encourage the group to continue their research.

Kornberg then discovered that when a bit of DNA extract from *E. coli* was added to the mixture, the real DNA acted as a primer or model for the test-tube "DNA." He also found out that the test-tube "DNA" was always a copy of the primer DNA even though the enzyme came from *E. coli.* When he used cow DNA as a primer, the enzyme made cow DNA; virus DNA as a primer resulted in virus DNA; primer DNA from another kind of bacterium fashioned DNA native to that organism. The primer DNA, regardless of its source, served as a template or mold which DNA polymerase copied.

The test-tube "DNA" synthesized by the Kornberg enzyme appeared to be an exact replica of real DNA chemically and physically. However, it lacked one property—it was not alive; it could not reproduce. Nevertheless, the test-tube "DNA" was close enough to real DNA to earn Kornberg a share of the 1959 Nobel prize in medicine or physiology.

The other 1959 Nobel recipient was Severo Ochoa of New York University, who was recognized for the test-tube synthesis of RNA. The crucial discovery here was an enzyme also extacted from a bacterium that linked together the RNA bases (A, C, G, and U). Ochoa's enzyme, it was later discovered, is not the enzyme that normally directs RNA synthesis in a cell, RNA polymerase. This is the enzyme that does for RNA synthesis what DNA polymerase does for DNA synthesis and requires a primer from which it copies RNA molecules.

In the Nobel prize citation, Kornberg and Ochoa (presented in alphabetical order) were described as "two of the best biochemists of the present time" and were recognized for "discoveries of the mechanism in the biological synthesis of DNA and RNA. These substances are thought to play a central role in transmitting genetic information from generation to generation and directing the synthesis of protein molecules in living cells."

Kornberg continued his attempt to create "living" DNA. Finally, in 1967, 12 years after the discovery of DNA polymerase, he announced that "living" DNA had been created for the first time outside a cell by Mehran Goulian, Robert L. Sinsheimer, and Kornberg at Stanford University. This achievement was not unexpected in view of previous research.

The DNA these scientists constructed was unusual in several ways. Instead of being the twisted double-stranded thread found in practically all living things, they made single-stranded DNA in the form of a ring consisting of 6000 bases making up nine genes. Single-stranded circular DNA is normally present in a few phages, including Phi X 174, the organism whose DNA was fabricated in this experiment. When this phage infects its host, *E. coli*, the injected single strand of DNA becomes the template and directs the formation of a second loop of DNA. The two loops intertwine to form a two-threaded ring. One loop then becomes the template for the formation of a new single strand, an identical copy of the original infecting strand. The new single strand becomes part of the new virus.

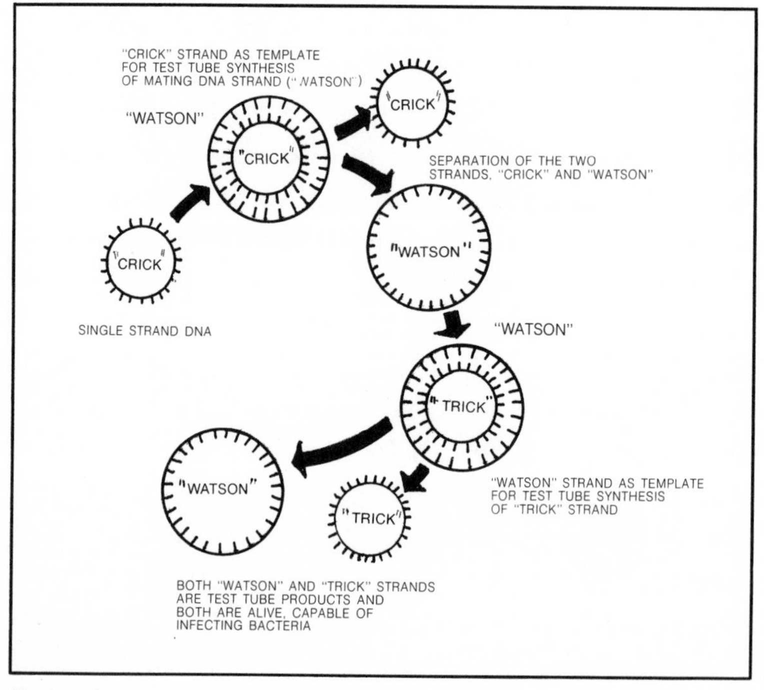

Test-tube DNA — from "Crick" to "Trick."

Basically, the process of duplication in single-strand DNA is the same as in the double helix in that complementary pairs of bases (A-T, and C-G) attract one another. The difference is that in the single-strand DNA only one strand functions as the template for making viral DNA. It is the original strand from which the duplicate strand is made.

The exact copy of Phi X 174 DNA, synthesized in a test tube by Kornberg and his colleagues, could do everything that natural viral DNA can do, leaving no doubt that it was "alive," capable of infecting *E. coli.* The successful synthesis opened many doors for further investigation of how DNA duplication takes place, a process far more complicated than it appears. This research has raised more questions than it has answered but has given us a more accurate picture of DNA duplication.

DNA DUPLICATION TODAY: DOUBLE AND SINGLE RINGS AND RODS

Until recently it was thought that all DNA molecules were double-stranded threads, that is, intertwining rods with two free ends. We now know that in several groups of bacteria and viruses the DNA is not only single stranded but also ring shaped. *E. coli* and Phi X 174 carry single-stranded ring chromosomes. Furthermore, the long, unbranched threads of DNA are condensed by being wound up like a spring. Circular DNA molecules are often coiled up so tightly that they form supercoils and the chromosomes resemble a series of beads in a necklace. When the chromosome of *E. coli,* which contains about 4 million bases, is uncoiled, it is a thousand times the length of the bacterium.

Little is known about the arrangement of DNA in the chromosomes of cells with a nuclear membrane where proteins are combined with DNA. In 1974 Roger A. Kornberg (son of Arthur Kornberg), then working at Cambridge University in England, proposed a "beads-on-a-string" model of chromosomes in cells with a nuclear membrane. The model strongly resembles the "beads-on-a-necklace" arrangement seen in circular DNA molecules in bacteria. The beads, it is thought, consist of tightly packed coils of DNA associated with special proteins

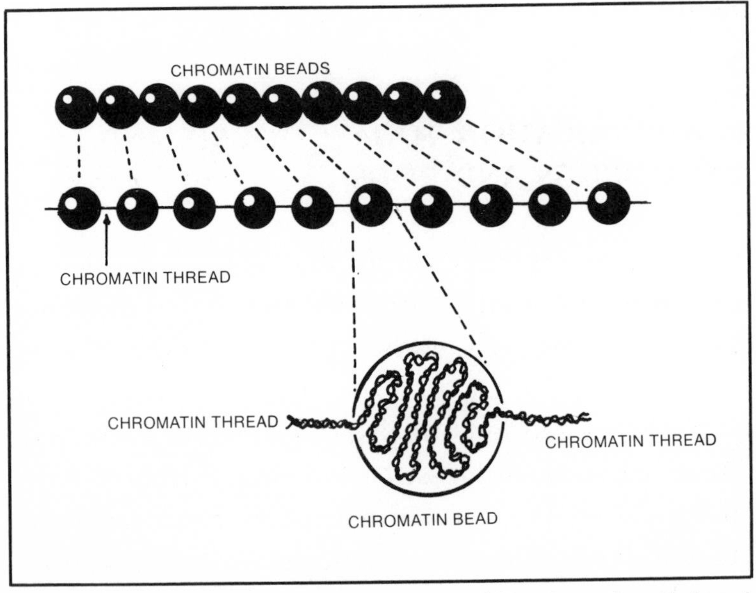

Chromatin threads resemble a string of beads and each bead contains about 200 bases.

called histones. Experimental data supporting this model have come from several sources, including work with frog cells in 1976 by Raymond Rieves and Alma Jones at the University of British Columbia in Vancouver, Canada.

UNWINDING AND WINDING DNA

The duplication of DNA starts with the unwinding and separation of the twisted strands of DNA. These gyrations are mediated by "unwinding" enzymes. Unwinding starts at a fixed point and the chromosome appears to contain a moving bubble between the strands. The bubble breaks out at a free end and assumes the shape of a two-pronged fork or a Y shape. As the strands separate the hydrogen bonds holding the bases together break and the bases are exposed. New chains of DNA begin to appear immediately on both sides of the fork next to the exposed bases. They are put together by DNA polymerase with incredible speed and accuracy.

ONE-WAY SYNTHESIS OF DNA

Arthur Kornberg found that the synthesis of the new chains goes in one direction only, in the 5′-to-3′ direction. These numbers refer to the position of the submolecules in the backbone of DNA, which consists of alternating phosphate and sugar units. The 3′ end of one strand is opposite the 5′ end of the other

One-way DNA duplication: The A. Kornberg model of DNA duplication assumes that the two DNA strands are duplicated by a single enzyme that joins together small pieces made in the 5′ to 3′ direction only.

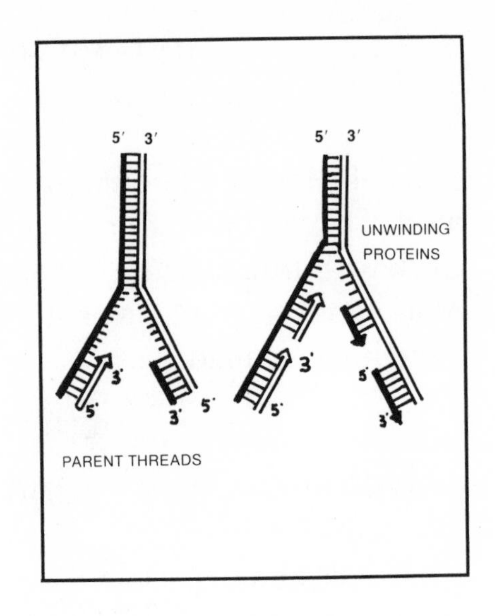

strand. The two strands lie in a "head-to-tail" position. This means that DNA polymerase moves up one leg of the fork and down the other. Kornberg proposed this explanation of DNA duplication by a single enzyme which can make DNA chains in the 5′-to-3′ direction only. He also suggested that the enzyme makes small pieces of DNA which are then joined together by another enzyme. A few months later the "joining" enzyme was isolated by several investigators, and at about the same time Reiji Okazaki and his colleagues at Nagoyo University in Japan found the predicted DNA fragments.

DNA-SYNTHESIZING ENZYMES

The complexity of DNA duplication is reflected in the number of different kinds of enzymes needed. At least a dozen are involved in the duplication of the *E. coli* chromosome. There are three known kinds of DNA polymerase. The Kornberg enzyme, now known as Poly I because it was the first discovered, has been found to have a repair rather than a duplication function. It fills in gaps and breaks in the DNA chain. The most recently isolated DNA polymerase, Poly III, is the main chain builder. It was discovered in 1971 at Columbia University by Malcolm L. Gefter and Thomas Kornberg, the other son of Arthur Kornberg. No known special function has been assigned to Poly II.

All DNA polymerases have a built-in self-correcting feature. They can "recognize" an incorrectly matched base, remove it, and replace it with the correct one. It is this ability that explains why so few "errors" are made during the duplication of thousands of bases in the DNA chain.

However, none of these enzymes can start a DNA chain. Instead, a form of RNA polymerase makes a small chain of RNA which serve as a primer for Poly III. Once started, Poly I removes the RNA primer and it is replaced on the DNA template by a fragment of DNA.

Damaged DNA is repaired by a group of enzymes working together in sequence. One kind of enzyme "recognizes" the damaged area and cuts through the backbone of the chain, a third enzyme cuts out the damaged portion, a fourth inserts the proper base, and a fifth joins it to the chain and the repair is complete. The precise function of the second enzyme is not known.

SUMMARY

DNA is a unique molecule capable of self-duplication. It usually consists of a double strand. Duplication occurs when the two strands separate and expose the bases, which act as templates for the synthesis of the two complementary strands. Duplication and separation go on simultaneously. The process is extremely complex; each step of DNA duplication is directed and controlled by an array of specific enzymes. Most of these enzymes have been isolated and employed in making test-tube "DNA." Duplication begins at a fixed point; the threads unwind and separate producing a fork along which replication takes place. The chain grows in one direction only, by adding small pieces which are linked together on the template strands. There are enzymes to unwind the strands, to start duplication, to elongate the chain, to terminate it, to proofread and correct errors, in base selection, and to fill gaps and repair damage to the chain. Most of what is known about DNA duplication comes from studies of bacteria and viruses.

DNA has two functions. The first is replication, which is extremely complicated and designed to minimize "copy" errors. The second is control and direction of protein production in the cell. Proteins are the workhorses of life. How they are produced and function is discussed in the next chapter.

PROTEINS UNLIMITED

DNA works its wonders by commanding the making of molecules. Most important are molecules of protein. DNA contains detailed instructions which are sent to the cell for making a rich assortment of protein molecules. DNA is the expert designer of proteins and the cell is the expert producer of them.

Proteins are huge, complex molecules essential for life. The very word *protein* means "primary, holding first place," which is precisely the position these molecules occupy in the scheme of life. All living things, from the smallest to the largest, contain these vital molecules. They are the major molecules in the structure of the cell. The membrane, cytoplasm, and nucleus are composed of proteins.

Every living part of your body has some kind of protein in it. Half the dry weight of your body is made up of tens of thousands of different proteins. They are part of your skin, hair, nails, muscles, tendons, ligaments, bones, and blood. Proteins are the chief molecular building blocks of life.

PROTEINS AT WORK

Enzymes are proteins that trigger practically every chemical reaction associated with life. Teams of enzymes digest your food, oxidize it, capture and transfer energy, and assist in building molecules such as DNA, RNA, and other proteins. An individual cell may contain as many as 1000 different enzymes, each promoting one of the hundreds of high-speed chemical reactions taking place in every cell of your body each second of your life.

The chemical activities of living things are coordinated by another group of proteins, the hormones (HORE-moans). In animals, these chemical coordinators are manufactured by special glands which secrete their hormones directly into the bloodstream. For example, the thyroid (THIGH-royd) gland in your neck produces the hormone thyroxin (thigh-ROCK-sin), which regulates the rate of oxidation in your body, that is, how fast the "flame" of life burns. The pituitary (pit-TYOU-ih-ter-ee) gland located at the base of the brain makes about 20 different hormones, many of which regulate the other hormone-producing glands.

Proteins known as antibodies are the chemical defenders of the body. They guard you against germs and the diseases germs cause. The polio virus stimulates your body to make specific polio antibodies which protect you against contracting this disease only. Organisms have an almost limitless capacity to produce antibodies, each specific for a particular germ or its poison. In general, proteins determine our form and function, that is, what we are and what we do.

MAMMOTH MOLECULES

Proteins are among the largest and most complex molecules found in living things. These molecules always contain the chemical elements carbon, oxygen, hydrogen, and nitrogen. Sulfur is usually present and there are often other elements. Compared to most molecules, proteins are enormous. Hemoglobin, the oxygen-carrying molecule in your blood, is an average-size protein containing 10,000 atoms. Of these 3000 are carbon, 5000 hydrogen, close to 900 oxygen, approximately 800 nitrogen, 8 sulfur, and 4 iron. Insulin, the hormone that controls sugar metabolism in your body, is comparatively a tiny molecule, with "only" 777 atoms.

PROTEIN BUILDING BLOCKS—AMINO ACIDS

Proteins are composed of chains of submolecular units called amino acids. There may be as few as 8 or as many as 100,000 of these building blocks in a protein molecule. Twenty kinds of amino acids occur commonly in proteins and they are used repeatedly as structural units. Hence, instead of six kinds of building blocks—sugar, phosphate, and four bases—as in DNA, protein may have as many as 20 varieties of amino acids, which may appear in innumerable combinations. The tiny insulin molecule contains 51 submolecular units, but there are only 17

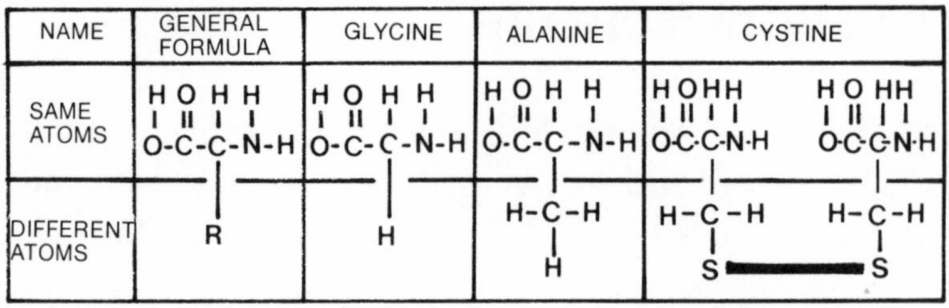

NAME	GENERAL FORMULA	GLYCINE	ALANINE	CYSTINE
SAME ATOMS	H O H H I II I I O-C-C-N-H	H O H H I II I I O-C-C-N-H	H O H H I II I I O-C-C-N-H	H O H H H O H H I II I I I II I I O-C-C-N-H O-C-C-N-H
DIFFERENT ATOMS	| R	| H	| H-C-H | H	| | H-C-H H-C-H | | S —— S

Chemical structure of amino acids: There are twenty kinds of amino acids commonly found in proteins and they are very much alike in chemical structure.

different kinds of amino acids included. Hemoglobin, the average-size protein, consists of 600 amino acids which fall into 19 varieties.

Amino acids are very much alike in chemical structure. They all have the same atomic backbone, a short chain of two carbon atoms and one nitrogen atom

$$-\overset{|}{\underset{|}{C}}-\overset{|}{\underset{|}{C}}-\overset{|}{N}-.$$

Each amino acid can be thought of as a railroad car with a pair of couplings, one in front and one in the rear. Just as railroad cars are hooked on to another to make a train, so amino acids are linked end to end to make long chains of these units

$$-\overset{|}{\underset{|}{C}}-\overset{|}{\underset{|}{C}}-\overset{|}{N}--\overset{|}{\underset{|}{C}}-\overset{|}{\underset{|}{C}}-\overset{|}{N}--\overset{|}{\underset{|}{C}}-\overset{|}{\underset{|}{C}}-\overset{|}{N}-.$$

The nitrogen atom of one amino acid can be hooked on to the carbon of the neighboring unit.

AMINO ACIDS IN CHAINS

Coming back to the example of the railroad train, any kinds of car—a flatcar, boxcar, tank car, passenger car, or caboose—can be coupled with another car and cars can be joined in any order. Similarly, any number of amino acids can be linked, and in any order. Each arrangement of these amino acids gives you a different protein.

How many different combinations of amino acids are possible in a protein? The number of kinds of proteins that can be made from the 20 kinds of amino acids is beyond belief. This becomes more evident if you think of the amino acids as letter s and the proteins as the words in this 20-letter alphabet. You have only to thumb through the pages of a dictionary to appreciate the number of English words that can be made with a 26-letter alphabet, and most of these are relatively short words with less than 10 letters. Now add the words in the French, German, Spanish, and Swedish dictionaries and you will begin to have some idea of the number of proteins that can be formed.

Amino acids are linked end to end.

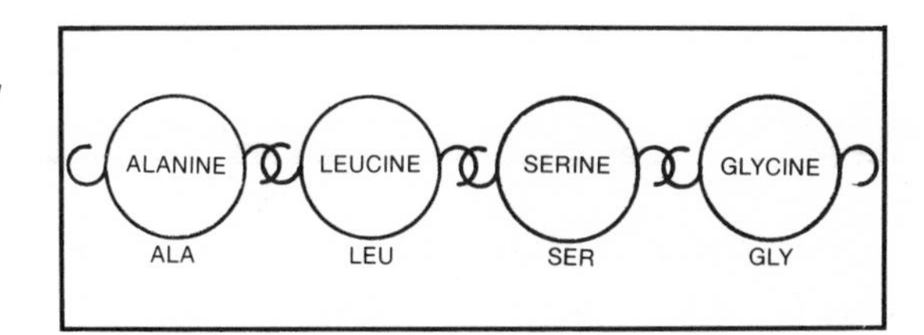

In the chemical language of proteins, the letters may be in any order and the words of any length. The average "word" contains 300 of these "letters." For example, a chain of just 5 different amino acids can be arranged in 120 different combinations. By doubling the number of amino acids in the chain to 10, the combination possibilities are increased to $3^1/_2$ million. In a chain of 20 amino acids, one of each kind, the number of possible different sequences is greater than the total number of living organisms on earth. How many possible combinations do you think there are in the average protein of 500 amino acids? The number is greater than the total number of atoms in the universe.

PROTEIN PROFILES

What is the structure of protein molecules? Scientists are interested in the architecture of these molecules, since they believe that what proteins can do depends upon how they are built. Each kind of protein seems to be especially designed for its particular task. A protein may consist of a single twisted chain of amino acids, or a bundle of twisted chains, or fibers bound together. To illustrate, the insulin molecule contains two linked chains, a long one with 30 amino acids and a short one with 21. Ligaments which hold together the bones of the body, like tendons which attach muscles to the bones, consist of bundles of coiled fibers bound

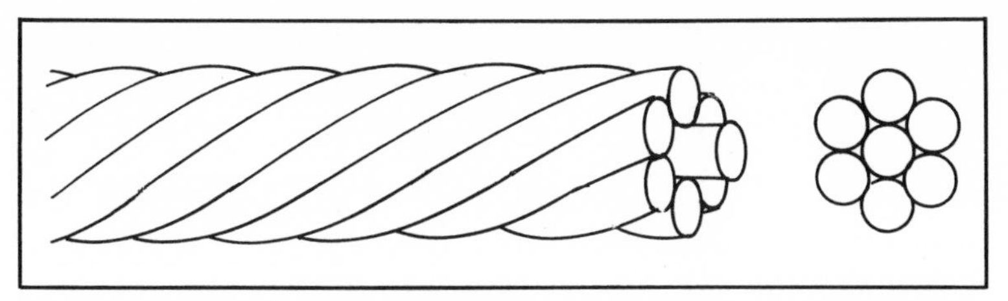

A protein may consist of a bundle of fibers resembling a cable. Each fiber has a spiral structure.

together like cable wires. There are some proteins in which the chains are folded back and forth into loops and coils assuming the general shape of a ball. Many hormones, as well as hemoglobin and enzymes, have this globular shape.

The first real breakthrough in deciphering the general shape of a protein molecule was announced in 1950 by Linus Pauling and Robert B. Corey at the California Institute of Technology. Using X-ray diffractions, they located the exact position and calculated the precise distance between the atoms in an amino acid molecule. They came to the conclusion that the amino acids do not follow one another in a straight line, but are twisted into spiral or coiled chains. For tracking down the shape of a protein molecule, Pauling was awarded the Nobel prize in chemistry in 1954.

AMINO ACID SEQUENCE

The second important advance was determining the exact sequence of the amino acids in the protein chain. This feat was first achieved at Cambridge University in England by Frederick Sanger and his

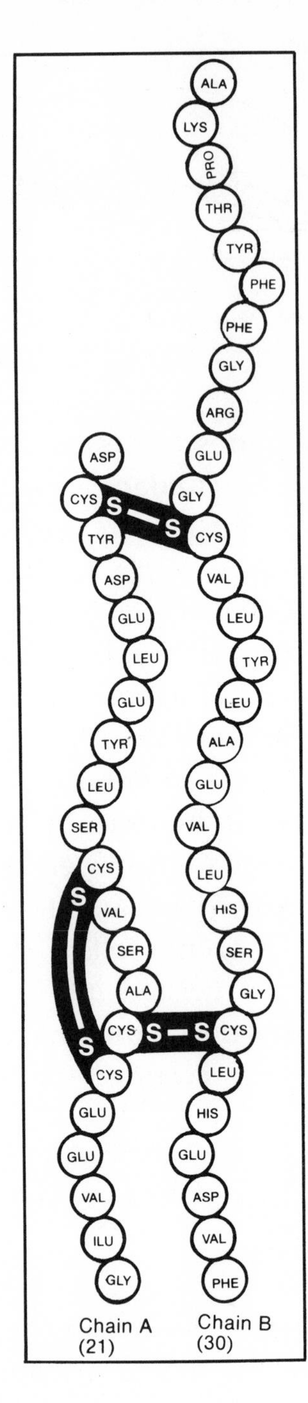

The insulin molecule consists of 51 amino acids arranged in two linked chains.

associates, who spent ten years studying the "tiny" insulin molecule. Not only did he discover the general shape of this molecule but, more important, he worked out the precise sequence of the 51 amino acids in its two twisted chains. For disclosing for the first time the internal structure of a protein—and proteins have been described as the "most complicated of all substances occuring in nature"—Sanger received the Nobel prize in chemistry in 1958.

Sanger's method was to break the insulin molecule into small fragments, to identify the amino acids in these pieces, and finally to put them together in their proper order like the pieces of a jigsaw puzzle. His success rested in part on a new technique, paper chromatography (crow-ma-tog-rah-fee), a method for separating the tiny pieces of insulin on filter paper and identifying the amino acids in them. A drop of ink placed on blotting paper forms a series of concentric rings of different colors as the different constituents of the ink separate out. This is a very simple example of paper chromatography. Try it on a few colored materials such as ink or beet juice.

In 1964 the first man-made protein was produced. A group of scientists at the University of Pittsburgh announced the synthesis of insulin.

ENZYMES REVEALED

Sanger showed the way for biochemical attacks on larger and more complicated protein molecules, the enzymes. Stanford Moore and William H. Stein of the Rockefeller University, together with Christian B. Anfinsen of the National Institute of Health, tackled the enzyme ribonuclease (rye-bow-NEW-klee-ace), the protein that breaks down RNA. With the help of automatic amino acid analyzers which Stein and Moore developed, they found in 1959 that ribonuclease consists of a single chain of 124 amino acids. Anfinsen figured out how the ribonuclease molecule develops its three-dimensional structure as the result of interaction among the various amino acids assembled in proper sequence. The amino acid chain is coiled into its three-dimensional structure by four cross-connecting bridges. For their "fundamental contributions to enzyme chemistry" this trio shared the 1972 Nobel prize in chemistry.

ANTIBODIES EXPOSED

1972 was the year of the protein molecule. The Nobel prize in medicine or physiology went to Gerald M. Edelman of Rockefeller University in New York and Rodney R. Porter of Oxford University in England for "discoveries concerning the chemical nature of antibodies," which are even more complicated protein molecules than enzymes. Working independently, Edelman and Porter unraveled the structure of a commonly found antibody, popularly known as gamma globulin and technically called immunoglobulin (IgG). This is a giant molecule containing 1320 amino acids in four chains arranged

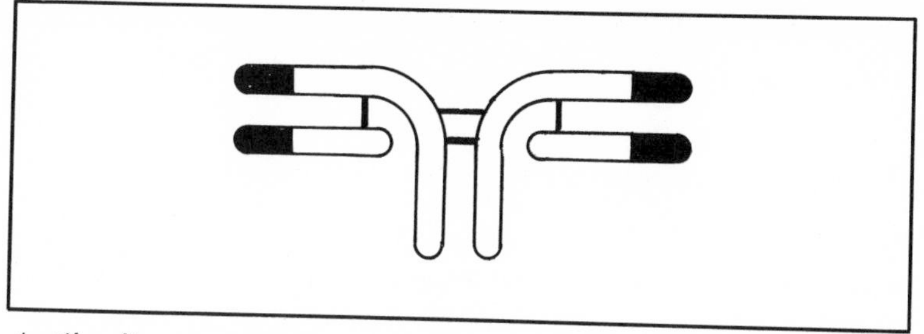

Antibodies are proteins composed of four chains of amino acids. The shaded ends of the chains are the variable portions of the molecule which are active in "fighting" disease. The rest of the chains are similar in composition and are inactive.

in the shape of a T. It consists of two long chains each with 440 amino acids and two short chains each with 220 amino acids. The two long chains form a pair of inverted L's side by side. Next to each of the two extending horizontal arms is a short chain parallel to it.

In all antibodies studied a portion of each of the four chains has the same amino acid sequence; however, the portion near the end of the arms has a variable sequence. It is thought that the variable portion, which is different in amino acid sequence and 3-D shape in each kind of antibody, is selected by a germ or its poison. This may explain why a given antibody acts only against a specific germ or its poison.

PROTEINS IN THREE-DIMENSIONS

The third and most recent front along which progress has been made in obtaining a complete picture of a protein molecule and how it works is its three-dimensional structure. The proper functioning of a protein molecule depends not only upon the number and kind of amino acids it contains but also upon their sequence in the chain. The way in which the amino acid chains are bent and folded three-dimensionally is determined by firing X rays at them to various depths and obtaining a series of "shadow pictures." When these pictures are stacked one on top of one another, they give the precise position of the atoms in the molecule.

Just about when Sanger began his work with insulin, another group of British scientists led by Max F. Perutz and John C. Kendrew, both of Cambridge University, began their study of protein molecules. Perutz aimed his X rays at hemoglobin (HEEM-oh-glow-bin), the oxygen-carrying protein in red blood cells; and Kendrew bombarded myoglobin (MY-oh-glow-bin), an oxygen-carrying protein present in muscles. After almost a quarter of a century, thousands and thousands of X-ray pictures, and mountains of mathematical calculations, the precise position of the thousands of atoms in these two protein molecules was pinpointed. Perutz and Kendrew found many similarities between these two very complex but functionally related protein molecules. With their data they constructed the first three-dimensional scale models of these molecules. Hemoglobin contains some 10,000 atoms and a total of 574 amino acids in four similar chains. Each hemoglobin chain bears a great resemblance to myoglobin, which consists of a single chain of 151

The myoglobin molecule.

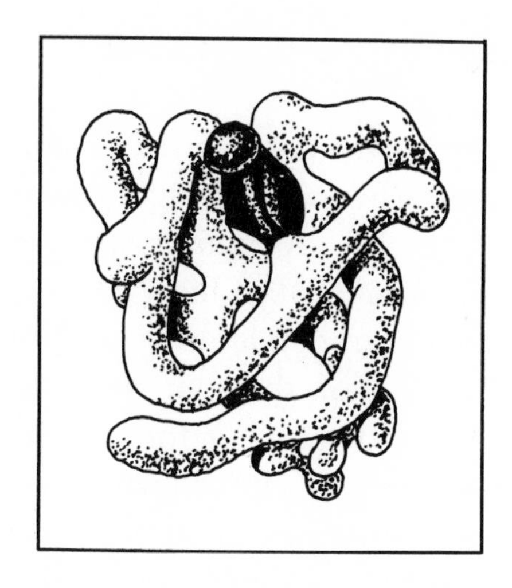

amino acids. Although the two proteins come from different animals and from different parts of the body—the myoglobin from the muscle of a sperm whale and the hemoglobin from the blood of a horse—these molecules are very similar. Each is a maze of interconnecting threads bent into intricate and elaborate three-dimensional patterns. But this was not the end of the surprise. They discovered that the threads in each of these molecules are twisted in the shape of a coil or helix. Thus two of the basic materials of life, DNA and proteins, are coiled threads—DNA usually a double thread and protein a single one.

In 1962, the year Crick, Watson, and Wilkins received the Nobel prize in medicine or physiology for their model of DNA, Kendrew and Perutz were awarded the Nobel prize in chemistry for their models of protein.

MORE X-RAY PICTURES

Improvements in laboratory techniques and the extensive use of high-powered computers for analyzing X-ray data have greatly hastened 3-D research. By 1965 the 3-D structure of an enzyme and some understanding of how it works was determined for the first time. David C. Philips and his associates at Oxford University completed their analysis of the enzyme lysozyme (LIE-so-zime), a protein found naturally in egg white and in human tears and other body fluids. Lysozyme protects the body against infection by dissolving the walls of certain bacteria. Its molecule consists of a single chain of 129 amino acids, part of it coiled into a spring, another part uncoiled, and a third part doubled back on itself to form loops. X-ray analysis shows that the molecule has two wings divided by a deep cleft. The cleft fits snugly over a sugar molecule in the bacterial wall and looks like a mouth biting into a slice of bread. Lysozyme appears to break the sugar molecule, puncturing the cell wall and so destroying the bacterium. The "jaws" of the enzyme close, bite the sugar molecule, open up, "spit out" the pieces, and are ready to devour more "sweets."

The 3-D designs of a dozen or so proteins are now known and we are beginning to understand how enzymes work. A very promising advance is the development of an optical computer capable of converting X-ray data almost directly into a 3-D

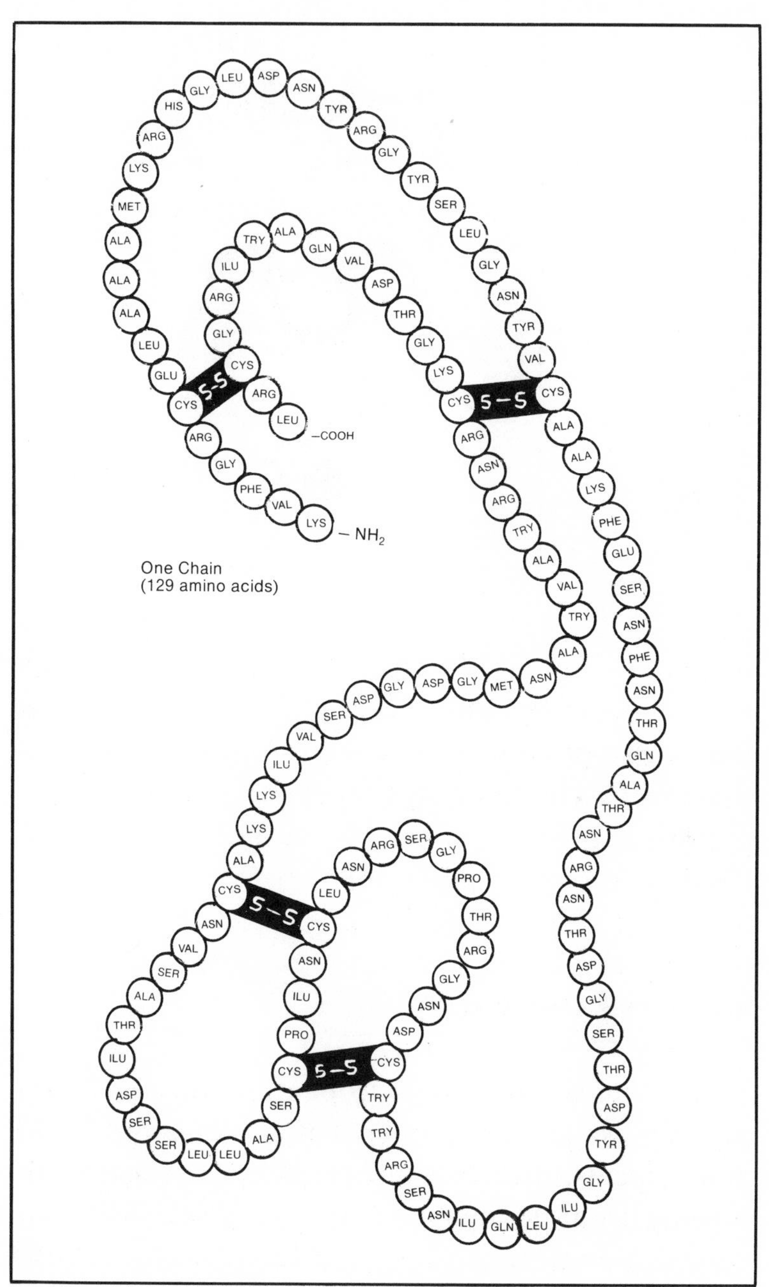

Lysozyme amino acids.

Lysozyme molecule in action—sugar molecule split.

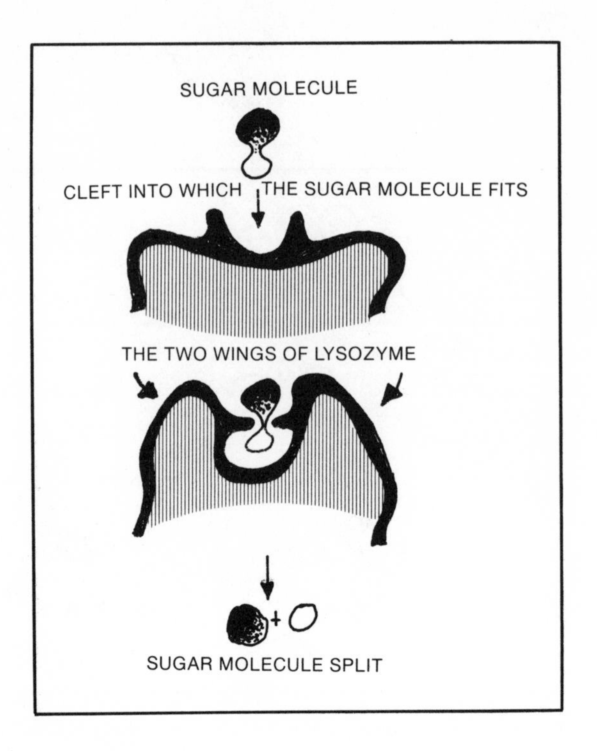

image of a molecule. George W. Stokes and his colleagues at the Stony Brook campus of the State University of New York are working on a method that reduces the time and effort required to produce a 3-D picture of a molecule by converting digital information into photographs.

PROTEINS UNLIMITED

Protein molecules differ from one another in the number, sequence, and spatial arrangement of the amino acids. The number of proteins that can be created from 20 amino acids is sufficient for every conceivable purpose in every living thing. But the kinds of proteins in each individual organism seem to be fixed and this is set by the genetic message.

DNA AND THE GENETIC CODE OF LIFE

The DNA molecule dictates the making of blond hair, or blue eyes, or any one of the thousands of traits that you have. How?

At this very moment scientists are trying to find answers to this question by "reading" the genetic information contained in the DNA molecule. They have "cracked" the genetic code and know the secret of heredity locked up in the DNA molecule. They are discovering how this molecule is coded to make a human or a horse, a fish or a fern, a tree or a toad.

We already know that DNA is a duplicating molecule that can make a copy of itself to pass on to the next generation. The reproductive cells, sperm and egg, which split off from the parent organism, contain parent DNA. At fertilization the microscopic sperm, containing the DNA of the male parent, enters the egg which holds the DNA of the female parent. The fertilized egg now contains packets of DNA from both parents and it is this combination of molecules that determines the heredity of the offspring. In this way, DNA molecules are transmitted from one generation to the next and the traits of both parents are inherited by the children. Bacteria also mate. Two bacteria move close to one another and a bridge forms between them. Through it, the DNA from one bacterium, the donor (or male), passes into the other bacterium, the recipient (or female). This is how bacteria mix DNA and pass it on to future generations.

BLUEPRINTS OF LIFE

The DNA in each organism contains thousands of specific messages which the cell "understands" and "obeys." The genetic messages are directions for making proteins. These vital molecules, you will recall, are composed of chains of amino acids, of which there are 20 common varieties. A typical protein contains a few hundred amino acids linked together in a specific sequence. DNA supplies the instructions to the cell for placing each amino acid in its proper place along the protein chain.

LIVING "PUNCH CARD"

The information in DNA is written in code that scientists have learned to "read" and understand. This code can be compared to the Morse code, the dot-dash language used in telegraphic communication. If I say to you "dash-dot-dot dash-dot dot-dash," it has no meaning unless you know the international Morse code. If you do, then you know the message reads "DNA." Similarly, the arrangement of the holes punched into cards fed into a computing machine means nothing to you. They do not give you any information. However, the number and the position of these holes are code symbols. When the cards are put into a computer, the pattern of these holes "tells" the machine what to do. The machine can "read" these symbols, follow the instructions, and translate them into words that you can read.

How is genetic information carried in DNA? The DNA molecules are the "punch cards" of life. The information they contain is in their molecular pattern just as the information on a punch card is in the pattern of the holes in the card. According to the Crick-Watson model, DNA has the shape of a twisted ladder with rungs containing only four combinations of base pairs: A-T, T-A, C-G, and G-C. The DNA in your body, in a cow, in a blade of grass, and in a virus, all has the same general molecular structure. The differences are in the *order* of the base pairs in the rungs of the ladder. The sequence of the bases along the DNA ladder spells out the genetic message and each species carries its own particular base sequence. In fact, except for identical twins, who are derived from the same fertilized egg, no two persons have the same order of bases in their DNA. This is another way of saying that each organism has its own particular genetic message which makes it different from all other living things.

THE FOUR-LETTER GENETIC ALPHABET

The symbols in the DNA molecule are the four bases found in the rungs of this ladder of life. Each base pair represents a different code letter. All the words and sentences contained in a genetic message are written in this four-letter alphabet of life.

Keeping the bases in a given order is as important in conveying genetic information as the correct sequence of letters in a word. If you shift the letters in CAT to ACT, the entire meaning of the word is changed. Altering the sequence of bases in DNA changes the genetic messages and the nature of the organism, perhaps from brown hair to no hair at all.

In a DNA molecule with just a few hundred base pairs, the number of different ways in which these pairs can be arranged is greater than the number of grains of sand in all the deserts of the world. The four-letter coded alphabet is sufficient for writing the fantastic amount of genetic information found in the 10 billion base pairs of the DNA in your cells. There seems to be no practical limitation to the number of ways in which these four symbols can be arranged to write out a different genetic story for each living thing. This is not too surprising when you remember that the Morse code has only two symbols, a dot and a dash, and that the computer also operates on only two symbols.

If you could unwind a coil of DNA, magnify it a few million times, and read it, you could probably identify the organism from the sequence of pairs of bases, which you will recall are adenine (A), guanine (G), thymine (T), and cytosine (C). This information would probably read like data on a ticker tape. Taking the names of the bases on only one side of the ladder, the beginning of the genetic code for one strand of a butterfly's DNA might read:

ACC TGC ATC.

That of a baboon might read:

ACC TTT CGG.

THREE-LETTER CODE WORDS: CODONS

DNA "speaks" only in words of three letters, that is, three bases in sequence. It sends messages to the cells consisting exclusively of words containing any three of the four letters in the genetic alphabet—A, C, T, and G. These code letters can be combined into 64 different three-letter code words or codons, such as:

AAA AAG AAT AAC

GAA GAG GAT GAC

CAA CAG CAT CAC

There are 52 other codons. How many of these can you make? The genetic code contains 64 different words and all hereditary messages are written in this language of life.

READING THE CODE

How should the genetic message be read? Scientists think that each one of the 20 different kinds of amino acids is coded by a three-letter word, and that a string of these words constitutes the genetic instructions for assembling a particular protein. For example, the message ACCAATAGAGGG is an instruction for making a protein containing four amino acids. The message should be read as a sentence containing four codons—ACC AAT AGA GGG. The first word ACC is the code word for one kind of amino acid *a,* the second word AAT for another amino acid *b,* the third word AGA for amino acid *c,* and the fourth word GGG for amino acid *d.* The decoded message reads: "Make a protein consisting of amino acids *a-b-c-d,* in that order."

CRACKING THE CODE

How did scientists find out which code word stood for which amino acid? How was the genetic code broken? A startling and spectacular breakthrough in cracking the genetic code came in 1961 when Marshall W. Nirenberg and J. Heinrich Matthaei, both working at the National Institute of Health in Bethesda, Maryland, successfully deciphered one of the 64 words. They made a molecule of RNA, a chemical cousin and close collaborator of DNA, consisting entirely of one of the four bases normally present in this molecule—uracil—which we shall represent by the letter U. This molecule, called Poly U, was put in a test tube with the protein-making machinery which they extracted from a living cell. They now added a mixture of all the amino acids. The protein-producing machinery set Poly U in operation and a protein turned up in the test tube consisting of only one kind of amino acid, phenylalanine (fen-ill-AL-uh-neen). Although all the other 19 amino acids were present and available, Poly U chose only the phenylalanine molecules and wove them into a protein chain. There appeared to be no doubt that the codon for phenylalanine was UUU. Thus, the first word had been placed in the genetic code book.

MORE CODONS

Poly U quickly became the key for uncovering more code words. Severo Ochoa and his colleagues at New York University, and Nirenberg and Matthaei, independently launched an all-out attack on the genetic code. They made RNA molecules containing every possible combination of A, G, U, and C, the bases found in RNA. In a little more than a year, a code word was found for each amino acid. In fact, they discovered that almost all of the amino acids have two or more codons. However, the letters in a codon must be arranged in their proper order to convey meaning. Genetic coding is like a game of anagrams—you may know all the letters but not the order in which they must be sequenced if they are to form words. If C, A, and T are the three letters in the codon for a particular amino acid, they can be arranged in six different ways: ACT, ATC, TAC, TCA, CTA, and CAT. Which is the correct spelling for the particular amino acid?

Can you read the coded message that follows?

Message:

CAA GAC AAA AAA ATA TTT CCC CAT
CAA AGG CAT ATA AAA AGA GAC

Code:

AAA—D	AGA—N	AGG—I
CCC—O	CAA—L	CAT—F
TTT—R	ATA—E	GAC—A

LEARNING THE CORRECT SPELLING OF CODONS

The breakthrough in discovering the correct spelling of codons is credited to Nirenberg and his colleague Philip Leder. In 1964 they announced that they found the correct spelling of the codon for the amino acid valine. There are three possible code words for valine, GUU, UGU, and UUG. They decided to synthesize each of these code words and find out which one correctly spells valine.

Prying the answer out of valine required the preparation of 20 batches of special molecules, each of which had the ability to pick out one of the 20 amino acids and bind it to a ribosome. These special molecules are called transfer RNA; the three-letter codons they made are tiny messenger RNA molecules. (All this is discussed in greater detail in the next chapter.) To each batch of transfer RNA molecules they added the proper amino acid plus ribosomes. Now came the crucial step, adding one of the three codons they had made and looking to see which of their synthetic molecules would glue which of the 20 amino acids to the ribosomes. This is the acid test, since it was discovered that the amino acid carried by its transfer RNA molecule will stick to the ribosome only when its codon is the same as that on the messenger molecule, in this case, on the synthetic codon. Nirenberg and Leder observed that of the three codons GUU, UGU, and UUG, only GUU glued the valine-carrying molecule to the ribosome.

The spelling code was cracked. The correct code word for valine is GUU. They later proved that UGU codes the amino acid cysteine and UUG spells leucine.

H. Gobind Khorana at the University of Wisconsin, using a slightly modified Nirenberg-Leder approach, confirmed the spelling of some codons and added more to the list. He and his students developed a method for making long messenger RNA molecules consisting of repeating codons—UCUCUCUC . . . — and examining the kinds of amino acids they assembled. Since this chain can be read as UCU or CUC, a chain containing two kinds of amino acids, serine and leucine, was made.

By the end of 1966, the test-tube techniques had yielded the code words for 61 of the possible amino acids. Three did not code for any specific amino acid but later proved to be "punctuation" codons that terminate a genetic message, like the period at the end of this sentence.

These results seemed to show that genetic messages consist of a series of three-letter codons which are read in sequence without gaps and without overlapping letters in the same way that you are now reading this sentence. For example, what does this following message say?

THEMANATETHEPIGANDTHEPEA

All this was so until recently when it was found that a virus message could be read in a different way and get different results. More about this later.

Teaching scientists how to read the genetic code rated special recognition, and in 1968 Nirenberg and Khorana shared the Nobel prize in medicine or physiology with Robert W. Holley of Cornell University "for discoveries concerning the interpretation of the genetic code and its function in protein synthesis." Holley made his mark with his studies of transfer RNA molecules. He worked out the base sequence of the transfer RNA molecule that picks up alanine, a task that required seven years of research.

THE GENETIC DICTIONARY—FIRST EDITION

The first edition of the completed genetic dictionary was published in 1968. The meaning of each of the 64 possible codons has been identified and verified, including 61 codons for the 20 amino acids and 3 codons for ending sentences in the genetic message. The codons are combinations of the four bases present in RNA—U, C, A, and G. In DNA, C, A, and G are present but T replaces U.

Each of the 64 codons is represented by three letters. The first comes from the left-hand column headed *First Letter,* the second is one of the four letters in the middle columns headed *Second Letter,* and the third is one of the four letters arranged vertically in the right-hand column headed *Third Letter.* For each of the 64 combinations an abbreviation of the amino acid for which it is coded is given: Phe for phenylalanine, Ser for serine, and so on. Phe

The Genetic Dictionary (1968 edition)

First Letter	Second Letter U	C	A	G	Third Letter
U	Phe	Ser	Tyr	Cys	U
	Phe	Ser	Tyr	Cys	C
	Leu	Ser	•Term	•Term	A
	Leu	Ser	•Term	Trp	G
C	Leu	Pro	His	Arg	U
	Leu	Pro	His	Arg	C
	Leu	Pro	GluN	Arg	A
	Leu	Pro	GluN	Arg	G
A	Ileu	Thr	AspN	Ser	U
	Ileu	Thr	AspN	Ser	C
	Ileu	Thr	Lys	Arg	A
	[s]Met	Thr	Lys	Arg	G
G	Val	Ala	Asp	Gly	U
	Val	Ala	Asp	Gly	C
	Val	Ala	Glu	Gly	A
	[s]Val	Ala	Glu	Gly	G

• Terminates the genetic message.
[s]Starts the genetic message.

is coded by two codons (UUU and UUC); Gly has four (GGU, GGC, GGA, and GGG).

The code also contains codons for starting and terminating a genetic message. The three codons UAA, UAG, and UGA are marked "Term" on the chart; they terminate the message like the dot at the end of this sentence. The beginning of the message, corresponding to a capital letter, is coded by AUG and GUG. AUG is also the code word for methionine (Met) and GUG is the codon for valine (Val). Why these codons are sometimes read as the coding for an amino acid and at other times as a "capital letter" is still a matter for scientific speculation and experimentation.

THE GENETIC CODE AS THE UNIVERSAL LANGUAGE OF LIFE

Is the genetic code the same for all living things or does the DNA in different organisms "speak" different languages? The question is being answered by finding out how the protein-making machinery in one species responds to the genetic message of another species. For example, UUU is the codeword for phenylalanine in mice and bacteria. Yeast and mold genes "work" in bacteria; rabbit genes for hemoglobin are obeyed in frog cells; and the tobacco mosaic virus genes are "understood" by bacteria. It seems that most of the genetic code may be universal, the same for all organisms—an assumption that holds until it is proved otherwise. The universality of the code may be looked upon as evidence of the unity of life at the molecular level.

THE DNA-RNA CONNECTION —FROM GENE TO PROTEIN

The way DNA controls protein production brings us back to the cell. The DNA designs for proteins are packed into chromosomes which remain inside the nucleus at all times. The *production* of proteins, however, takes place outside the nucleus, in the ribosomes, the protein-producing centers of the cell, situated in the cytoplasm. The raw materials—the amino acids—needed by the ribosomes for manufacturing proteins are supplied largely by the foods you eat. The enzymes in your digestive system split the proteins in your food into amino acids, which are eventually distributed to the cells of your body. There DNA organizes these food units into the many kinds of proteins needed by the body. Thus the DNA in your cells takes the amino acids which came from cow protein in the form of a steak, and reassembles them to make human proteins. What you eat turns into you. How is this done?

RNA—MOLECULAR MESSENGER OF DNA

How can DNA, which is confined to the nucleus, control a process going on in the cytoplasm? What is the link between DNA and the ribosome? The gap

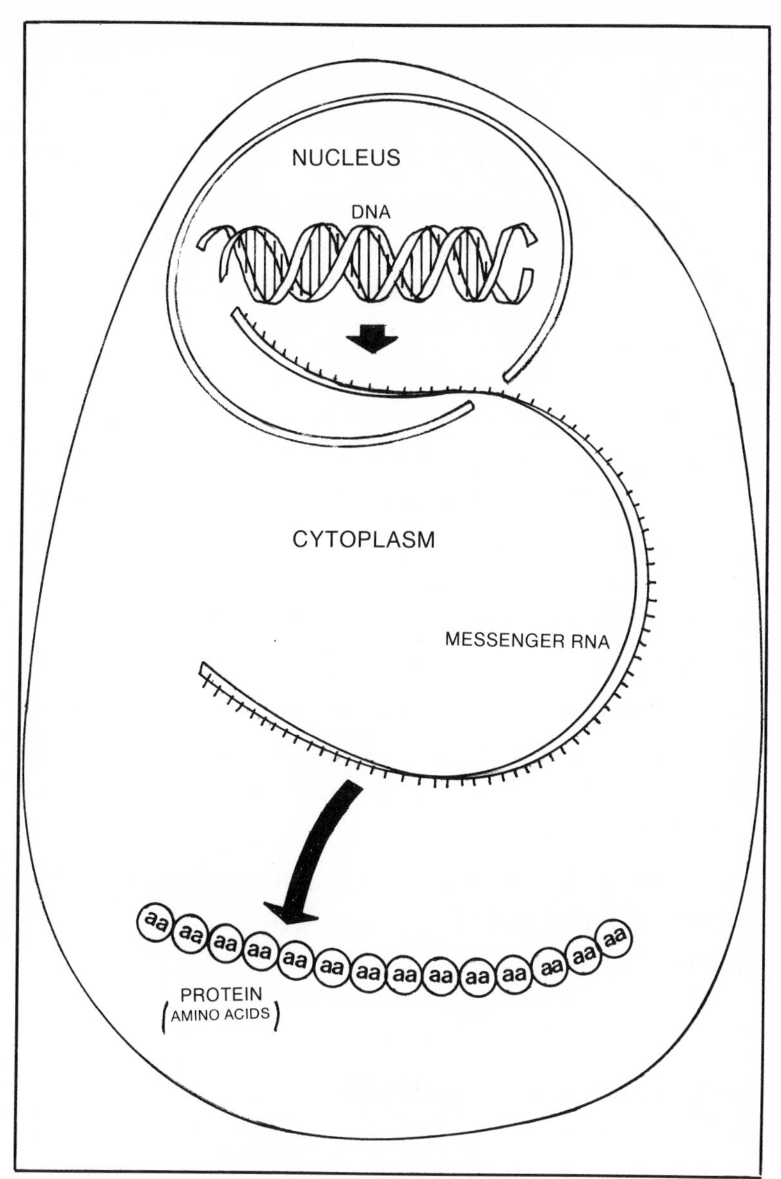

How DNA controls protein production.

between the plan and the protein is bridged by myriads of molecules of RNA, ribonucleic acid, which is a chemical first cousin of DNA. There is a continuous flow of instructions from DNA, the master planner, via RNA, the master contractor, to the ribosome, the master builder. RNA molecules transmit the genetic messages to the protein factories; they tell the ribosomes the kinds and quantities of proteins to produce. RNA is directly involved in the protein-making business of the cell.

There are many reasons for regarding RNA as the molecular messenger of DNA. In a cell, DNA is present only in the nucleus, whereas approximately 90% of the RNA is located in the ribosomes of the cytoplasm. In addition, delicate techniques have detected the nucleus in the act of discharging bits of RNA into the cytoplasm.

All cells contain RNA but some contain more than others. Cells which produce large amounts of proteins contain large amounts of RNA. Growing cells have more RNA than resting cells, because growth involves increasing their protein content. The cells of the pancreas and liver, which are constantly making and secreting proteins such as enzymes and hormones, are very rich in RNA.

RNA—HALF A DNA LADDER

An excellent reason for thinking that RNA carries out the instructions of DNA is the chemical and physical resemblances between them. Both are composed of six sub-molecular pieces. However, instead of looking like a twisted ladder with two strands connected by rungs, RNA generally resembles half of a twisted ladder. The backbone of this single-stranded mole-

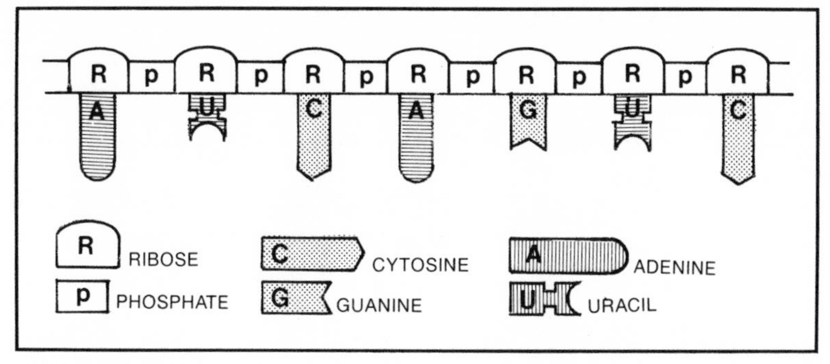

RNA molecule.

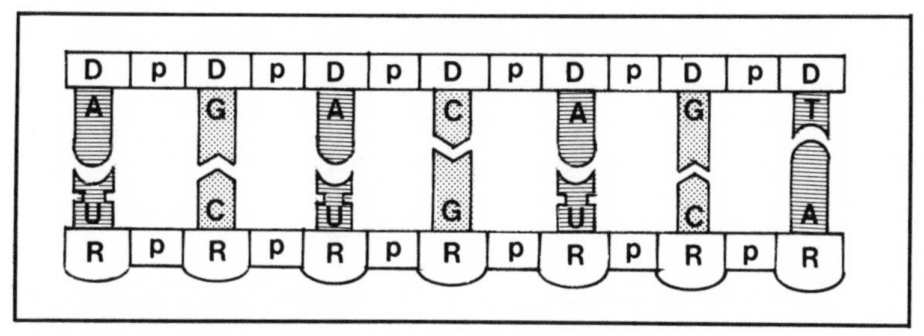

How DNA makes RNA.

cule consists of alternating sugar and phosphate units. The sugar in RNA, however, is ribose, which has one more oxygen atom in its molecule than the deoxyribose of DNA. There are four bases connected to the sugar units, three of which are identical with those found in DNA—adenine (A), guanine (G), and cytosine (C). The fourth RNA base is uracil (U) in place of the thymine (T) in DNA. Note that uracil has a close chemical resemblance to the thymine it replaces.

RNA chains vary in length: some contain as few as 75 bases while others may contain over 10,000. The chains take various forms; parts of long molecules are a double thread. As a chain loops back on itself in hairpin fashion, matching bases in one part

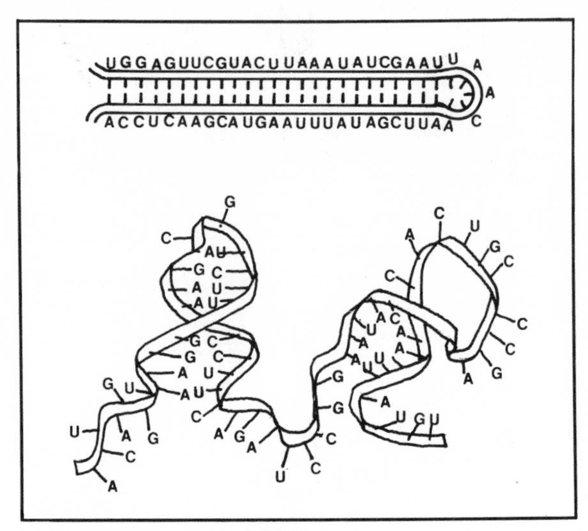

Folding patterns of RNA.

of the strand can bond with complementary bases in the other part (G with C and A with U) and appear as a partly double-threaded structure. No circular RNA chains have been found so far—a shape which DNA assumes in some bacteria and viruses.

MESSENGER RNA (mRNA)—THE CODE-CARRYING MOLECULE

One kind of RNA molecule copies the genetic message from DNA and carries it to the ribosome where it is "read" and translated into proteins. This code-carrying molecule is appropriately named messenger (mRNA). Each mRNA thread is a copy of the DNA instructions for making one or more proteins. There are as many kinds of mRNA molecules as there are kinds of proteins in an organism. mRNA molecules also vary considerably in length depending upon the size of the protein being "ordered" including the "start" and "stop" signals necessary. An order for an average-size protein of 400 amino acids requires a mRNA molecule of no less than 1500 bases. The mRNA for making the amino acid histidine contains coding for the 10 enzymes needed to make it and therefore consists of more than 10,000 bases. After leaving the chromosome, mRNA molecules attach to one or more ribosomes. Here they become the template on which the amino acids are assembled in the sequence dictated by the sequence of bases on the mRNA molecule. After the message has been read and the protein chain is completed, their job is done and the mRNA may disintegrate or be recycled.

RIBOSOMAL RNA (rRNA)—THE CODE-READING MOLECULE

The second DNA creation is ribosomal RNA (rRNA) molecules. These molecules together with ribosomal proteins constitute the ribosome, the miniature protein-producing factory. It is in the ribosome that

Ribosomes consist of two subunits, one twice as large as the other.

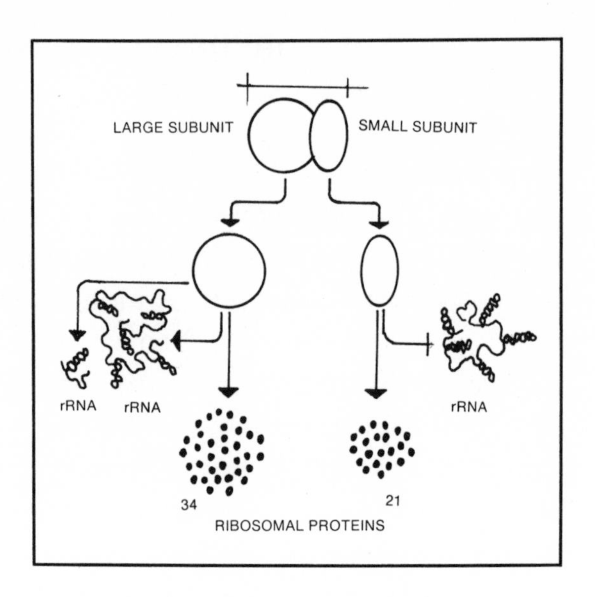

the genetic message is translated into proteins. Ribosomes occupy the central position in the passage of genetic information from the gene to the protein. Despite their importance we have only a glimmering of how ribosomes function. Most of what we know comes from studies of microbes such as *E. coli*, and its phages.

Ribosomes are constructed from two subunits, one twice as large as the other. Both contain RNA and proteins. The larger subunit contains 34 specific proteins and two different kinds of mRNA molecules; the smaller consists of 21 distinct proteins and a single rRNA molecule. The number of bases in each of the three kinds of mRNA molecules present in the ribosome is known and the complete base sequence of the smallest rRNA molecule has been determined. Note that there are three kinds of mRNA in the ribosome, two in the larger subunits and one in the smaller.

It is thought that the smaller ribosomal subunit is the site for amino acid assembly and the larger subunit provides the enzymes for hooking the acids together. Both subunits must be combined with mRNA for protein synthesis to occur. Since rRNA does not seem to carry a genetic message, its precise role in protein synthesis is not clear. Nothing is known about the 3-D structure of rRNA, let alone the ribosome.

mRNA can insert itself into several ribosomes successively and move across many ribosomal surfaces simultaneously, like a string being pulled through beads. As the ribosome traverses the mRNA strand it "reads" the coded message as a tape recorder reads a tape. The message calls for the ordering of certain amino acids in a specific sequence. Hence as mRNA moves through the ribosome, a chain of amino acids is being built by translating the base sequence in mRNA into the amino acid sequence of the protein for which it is coded. At the end of the synthesis the ribosomes drop off the mRNA strand, releasing the protein chain and separating into large and small subunits. Several ribosomes can read different parts of mRNA simultaneously and make several proteins. But this does not end the protein synthesis story.

TRANSFER RNA (tRNA)—THE AMINO ACID-CARRYING MOLECULE

The third DNA-made member of the protein-producing team is the transfer RNA (tRNA) molecules. These are engineered so that one end of the molecule can be attached to a specific amino acid while the other end sets that amino acid in its proper place on the mRNA template. There are no less than 20 kinds of amino acids and hence 20 kinds of tRNA, one for each kind of amino acid. tRNA for the amino acid alanine can not only "spot" alanine in a crowd of 20 different amino acids but can "latch" on to it and bring it to the alanine codon on mRNA. How does tRNA for alanine "recognize" alanine and also its codon on mRNA? The same question can be asked for the other 19 amino acid tRNA molecules. Answers to these and similar questions are now becoming available.

It was not until 1964 that the chemical structure of a tRNA molecule was elucidated. Robert Holley, one of the three 1968 Nobel prizewinners in medicine or physiology, completed the determination of the base sequence of the 77 bases found in tRNA for the amino acid alanine present in yeast. Since then the base sequence of practically all tRNA's has been determined. They are all similar in size, shape, and base sequence, regardless of source. tRNA molecules are relatively small, averaging about 80 bases arranged in a single chain. Parts of the chain have hairpin folds which are held together by the hydrogen bonding of complementary bases (A-U and C-G) and resemble the double helix of DNA. Unpaired portions form the loops and a single-threaded end stem. Two-dimensionally, tRNA molecules assume a

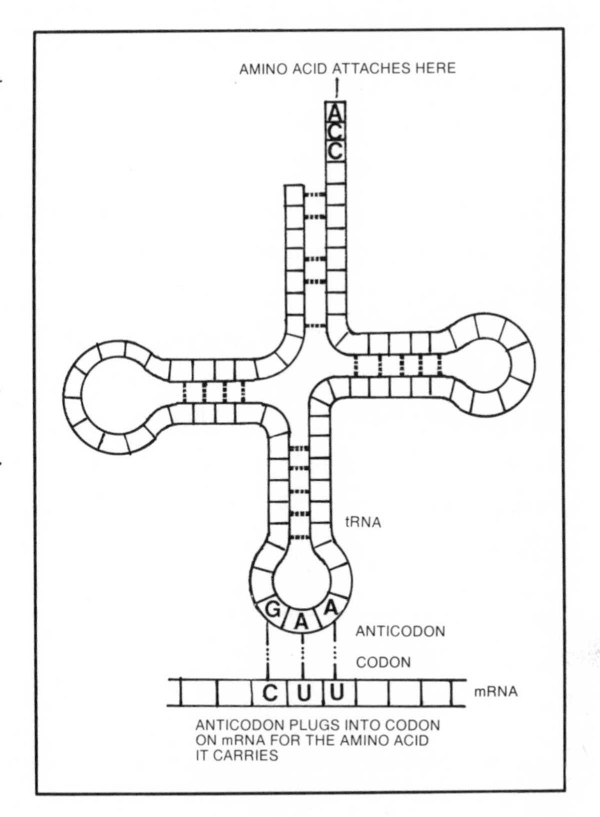

Transfer RNA molecules consist of a single chain of bases, parts of which have hairpin folds held together by the hydrogen bonding between complementary bases (A-U and C-G).

cloverleaf shape and are affectionately referred to by biologists as the "flower" molecules.

The end of the stem of the cloverleaf molecule always contains the same three unpaired bases: A, C, and G. The amino acid always attaches itself to the end base A. This is the amino acid acceptor part of the molecule. The loop at the opposite end contains three unpaired bases. This triplet is coded to match the codon on mRNA for the specific acid it is carrying, and is referred to as the anticodon loop. For example, the three bases in the tRNA anticodon loop for the amino acid proline are CGG. This tRNA molecule can bind to mRNA only where a complementary codon occurs, namely at GCC. In this way the proper placement of an amino acid is assured. Another interesting similarity is the distance between the amino acid acceptor and the anticodon loop ends; it is the same in all tRNA molecules. Perhaps this is the way to line up all the amino acids so that they will be next to each other on mRNA, facilitating their linkage into chains.

X-ray diffraction studies have added the third dimension to our knowledge of the structure of tRNA molecules and some insights as to how they

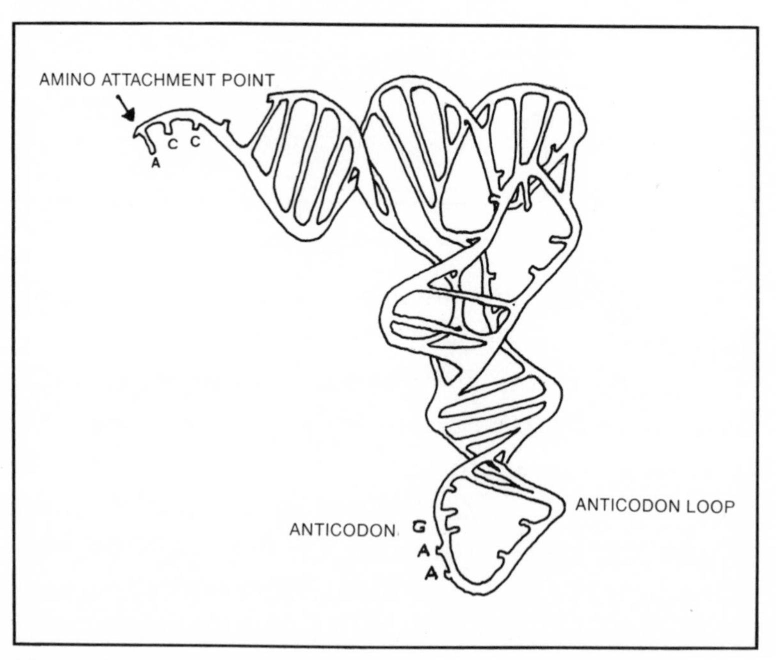

Three-dimensional shape of tRNA.

function. In 1974, under the leadership of Alexander Rich of the Massachusetts Institute of Technology an analysis of the 3-D structure of a tRNA molecule was completed and the exact position of all 77 bases was determined. The double-helix stems present in the two-dimensional model were verified. In addition, and quite unexpectedly, the team found several unusual additional hydrogen bonds among the bases, which bent the cloverleaf into an L-shaped structure. Since these bondings were between bases common to all tRNA molecules, it was deduced that all tRNA molecules have the same 3-D configuration. tRNA molecules display the same universality as the genetic code, again probably an indication of the unity of life at the molecular level.

MAKING RNA MOLECULES

How is genetic information for making proteins transferred from DNA to RNA? DNA makes a molecule in its own chemical image specifically for this purpose. A segment of the DNA molecule unwinds and one of its strands serves as a template or mold for lining up the proper RNA chemicals along its length. The process is much like that in DNA duplication except that as previously indicated, the chain being assembled contains uracils (U), in place of thymines (T), and ribose instead of deoxyribose. The

A, C, G, and T bases in the DNA template pick up U, G, C, and A respectively. The sequence of the bases in the DNA template dictate the order of the bases in the RNA thread being assembled. An RNA molecule is the chemical copy of the DNA molecule that makes it. Hence whatever instructions there are in DNA for a given protein molecule are stamped into the RNA. However, making proteins is no simple matter; it requires the help of several kinds of RNA molecules, all of which are DNA creations.

As the various places on the messenger RNA template are filled with the correct amino acids, they join with one another, forming a protein chain. The transfer RNA molecules are set free to ferry more of the same amino acids from the cytoplasm to the ribosomal template. The newly made protein molecule peels off the template and goes to work in the life stream of the cell.

The entire process takes place in a matter of seconds. Richard Schweet of the University of Kentucky and Howard M. Dintzis of the Massachusetts Institute of Technology found that the amino acids making up hemoglobin are joined together in the ribosome in zipperlike fashion. Starting at one end of the chain, amino acids link up one after the other until the entire protein molecule is complete. It takes less than 2 minutes to hook together the hemoglobin molecules consisting of four chains each with 140 amino acids, at the rate of 10 per second. In *E. coli* a chain of 300 to 400 amino acids is put together in 10 seconds.

FROM GENE TO PROTEIN—A SUMMARY

Thus the DNA molecules contain instructions for making proteins. The "blueprints" for each kind of protein are called genes. Each gene is located on a specific portion of DNA coil and the genes follow one another along the length of the DNA molecule like bits of information on a ticker tape.

DNA, the director of the cell, controls all cellular activities from the executive suite in the nucleus. The master plan for making thousands of proteins are filed in separate genes stored in the cell's filing cabinets, the chromosomes.

Being a good executive, DNA does not do all the work itself. It creates assistants and gives them responsibilities. These assistants are the various kinds of RNA molecules. One group of assistants consists of a messenger, RNA (mRNA) molecules, which are chemical copies of genes containing the blueprints for making specific proteins. These mRNA molecules move out of the nucleus into the cytoplasm and attach to ribosomes which house the second member of the team, rRNA. The precise function of rRNA and the ribosome is still unclear but their importance is undeniable. Ribosomes seem to coordinate the activities of mRNA and the third member of the protein-producing trio, tRNA. There are 20 kinds of tRNA molecules, one for each kind of amino acid. tRNA has two attachment points. The specific amino acid hooks on to one end of the molecule while the other end of the tRNA molecule

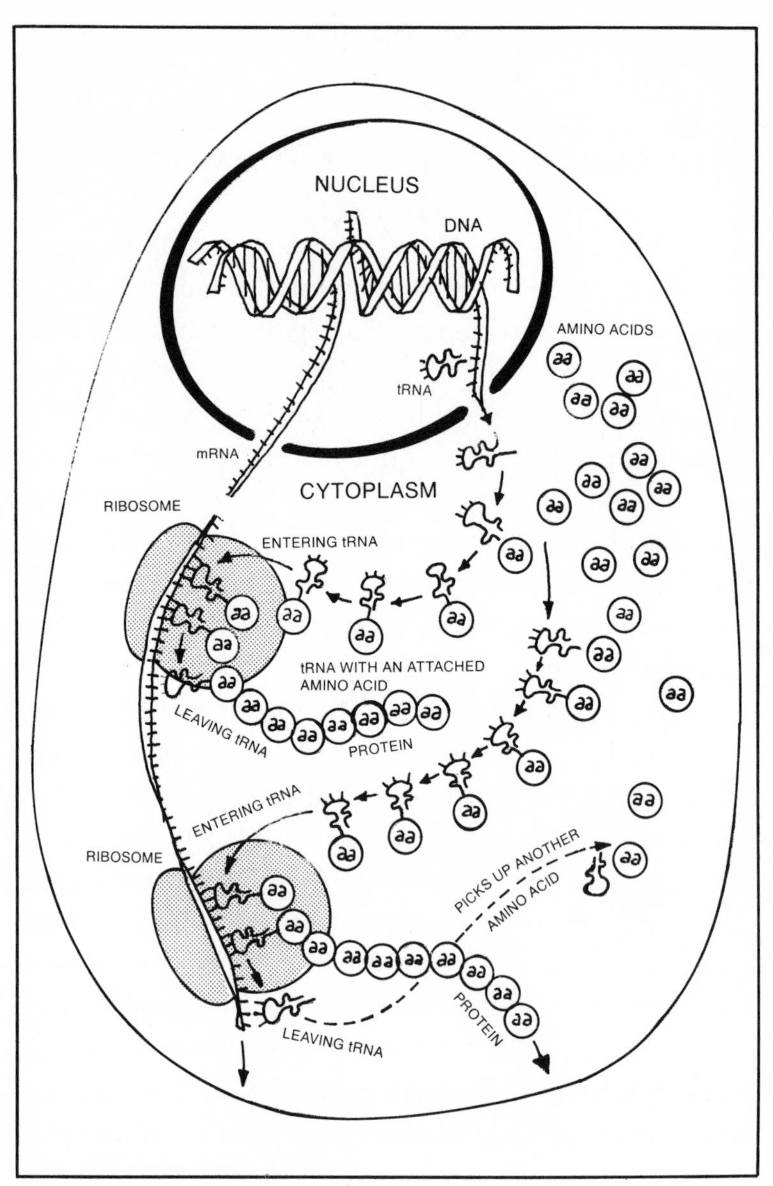

Summary of protein synthesis.

plugs into the proper place for that amino acid on the mRNA strip. The tRNA molecule finds its correct position by the three bases in the anticodon loop which pair with complementary bases for that amino acid on mRNA.

As the ribosomes move along the mRNA thread, the amino acids join together one by one, stepwise, to form a chain. As each tRNA surrenders its amino acid to an incoming tRNA molecule it detaches itself, and is free to repeat the cycle. When the end of the mRNA line is reached the newly formed chain of amino acids spontaneously folds into a protein molecule and slips off into the cytoplasm. At the same time the ribosome separates into its two subunits, freeing it either to make more of the same protein or available to receive a new set of instructions. The components of the protein-making machinery—rRNA and the ribosome, tRNA, mRNA, and an assortment of enzymes—turn out proteins with astonishing speed, accuracy, and efficiency.

The cycle from gene to protein is repeated over and over again until the death of DNA. Many of the stages of the DNA-RNA protein production partnership in bacteria have been observed using the electron microscope. But much of what we have seen and dreamed of is only the tip of the iceberg of knowledge of this most intricate and complicated process. This becomes more evident in the next chapter, which discusses the mechanisms that control DNA activities in the cell.

DNA CONTROLS—IN MICROBES

A living thing is a masterpiece of complex structures, organized and integrated to conduct the business of life. Some cells make blood, others produce bone, still others manufacture insulin. Each kind of cell "does its thing" but only when needed for as long as necessary and for the good of the organization. And yet each cell has a complete set of identical genes programmed for all the activities and structures of life. Obviously, all genes are not turned on at the same time. For example, a single cell from the inside of a carrot can grow into a complete carrot. A frog egg cell whose nucleus has been destroyed and replaced by the nucleus from the frog's intestines matures into a complete, perfectly normal frog. All the genetic information needed to grow, develop, and function as a complete frog is present in the nucleus of the intestinal cell. Something determines which genes act, when, and how often. In their daily activities, genes are constantly being turned on when needed and turned off when not needed.

Sets of genes may be compared to the strings of a harp. All the notes are there; any one or more of them can be played at any time in any combination or sequence. Each cell "plays" its own tune, which is orchestrated in the symphony of life.

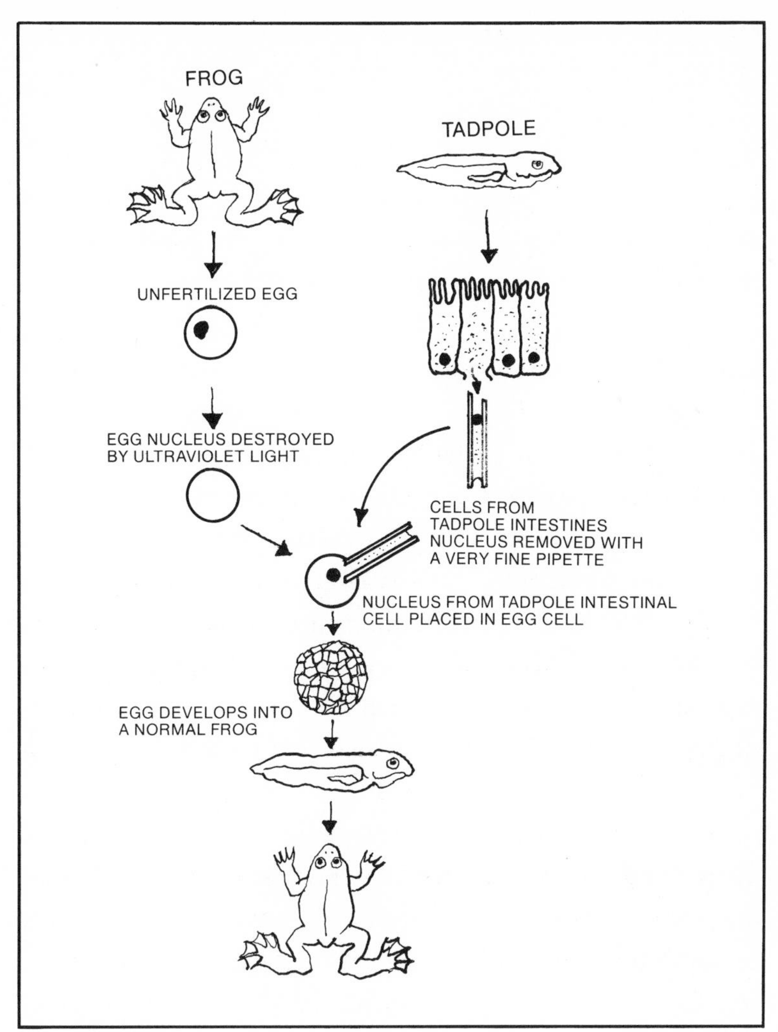

Cloned frog.

THE FEEDBACK MECHANISM

The basic questions are "How are genes controlled?" "How do genes know when to turn on and off?" and "How are genes regulated and coordinated?" A hint may be found in a feedback control mechanism. A simple example is the thermostat which controls room temperature. Too much heat opens the thermostat, cutting off the heat; too little heat closes the thermostat, turning on the heat again. Living things have a similar feedback mechanism which enables them to adjust to changes in the environment. Too much sugar in your blood stimulates the pancreas to make more insulin which enables the body to remove the excess; too little sugar causes you to reduce insulin production until the proper amount of sugar is present. When you eat starchy foods, your stomach and intestinal cells make and secrete starch-digesting enzymes. No starchy foods, no starch-digesting enzymes. Bacteria do the same thing we do. When *E. coli* bacteria are fed the milk sugar, lactose, they make lactose-digesting enzymes; no lactose, no lactose-digesting enzymes. The amount of lactose given to the bacterium also specifies the amount of enzymes made. These feedback systems are very efficient and economical. They make, deliver, and use materials according to the needs of the organism. Now let us see how the feedback mechanisms fit into the master control system of living things.

The road between genes and proteins is carefully monitored and controlled at several strategic points. The first place is on the DNA thread during the process of transcription when the genetic instructions on DNA are being transcribed to messenger RNA. The second place is on the ribosome where the message is being translated into a protein, that is, the base sequence of the mRNA is translated into the amino acid sequence of the protein. The third place is in the cell where proteins such as enzymes and hormones are performing their tasks. Controlling *transcription* means turning on certain genes and turning off others; controlling *translation* means determining how much of a particular protein is to be made; and controlling *the activities of the protein* means determining how well it is doing its task.

JACOB-MONOD MODEL—FRENCH FASHION

The first study which attempted to scientifically explain gene control was published in 1961 by two Frenchmen, François Jacob and Jacques Monod from the Pasteur Institute in Paris. They found that *E. coli* bacteria make lactose-digesting enzymes only when fed lactose. The lactose molecule is the essential ingredient in the feedback mechanism; it exercises control over the genes that make the enzymes by simply turning them on or off. The Jacob-Monod model presents an explanation of the molecular mechanisms involved.

Genetic studies indicate that there are three genes next to each other on the *E. coli* chromosome that make the lactose-digesting enzymes. These three genes are called structural genes because they determine the structure of these proteins. Jacob and Monod found a fourth gene located next to the three enzyme-producing genes which they called the operator. It controls the activities of the other three genes. The operator and the structural genes

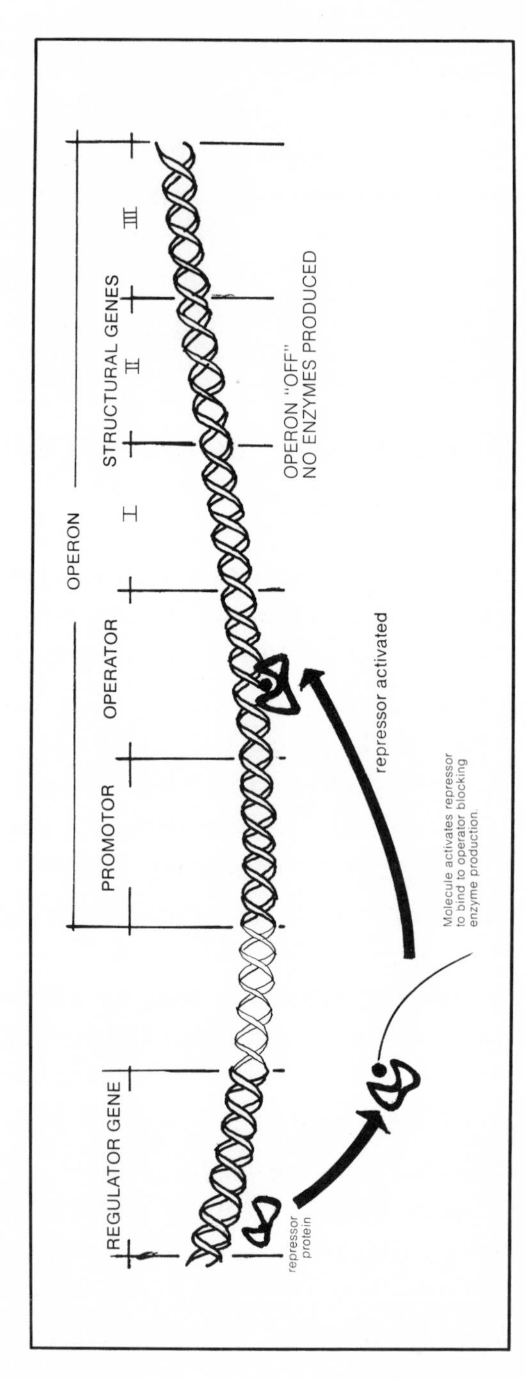

The Jacob-Monod model of gene control in bacteria.

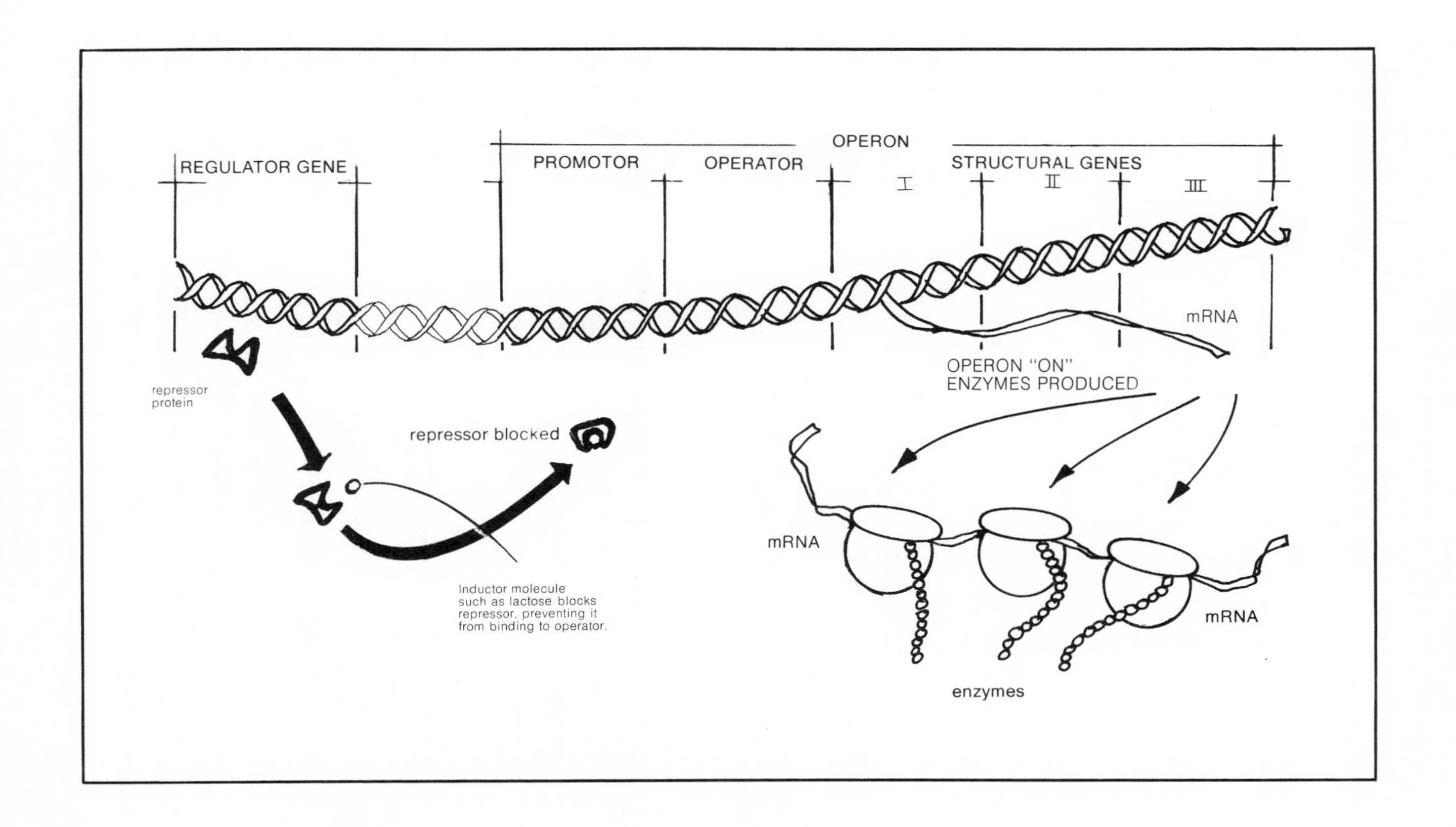
OPERON
REGULATOR GENE
PROMOTOR
OPERATOR
STRUCTURAL GENES
I
II
III
mRNA
repressor
protein
OPERON "ON"
ENZYMES PRODUCED
repressor blocked
Inductor molecule
such as lactose blocks
repressor, preventing it
from binding to operator.
mRNA
mRNA
enzymes

work together as a functional unit, which they named the operon. When the operator is "on," all three genes are working; when the operator is "off," the genes are not making enzymes. Thus the operon may be compared to a wall switch that operates three lamps in a room.

The investigators were faced with the problem of explaining how the operator gene is turned on and off. They concluded that a fifth gene is involved, located some distance from the operator gene; they labeled it the "regulator" gene. The regulator, they believed, produces a protein molecule which binds to the operator gene and keeps it in an off position. These molecules are called repressors since they repress the production of enzymes by the three structural genes.

The repressor is regulated by other molecules such as lactose. This molecule may combine with the repressor and prevent it from binding to the operator. The operon is "on" and produces enzymes. Other molecules activate the repressor enabling it to bind to the operator. This turns "off" the operon and no enzymes are produced.

THE FRENCH MODEL—A WINNER

The Jacob-Monod model has had a profound influence on molecular biology and DNA research. It is ranked as one of the more important milestones in the history of molecular biology comparable to the cracking of the genetic code and the discovery of messenger RNA. In 1965, these French model makers shared the Nobel prize in medicine or physiology with their colleague and countryman Andre Lwoff, who proved that the genes of a virus can become part of the genes of a bacterial host and that the host can duplicate both the viral and its own genes for several generations.

The French model of gene control resembles the Crick-Watson double-helix model of gene composition in several ways. Both were theoretical creations, the products of scientific imagination, which provided guidelines for further research and opened up new paths for investigation. Both have stood up fairly well over the years in the face of scientific scrutiny and testing.

The Jacob-Monod model assumed the existence of messenger RNA, which was later discovered. There was then no direct evidence for the repressor genes—these were also found later. On the other hand, research has revealed that gene control is far more complicated than as originally conceived by Jacob and Monod.

REPRESSOR SUBSTANCES AND SITES

Support for the French model came from the discovery and isolation of repressor substances in several bacteria and viruses. In 1966 Walter Gilbert and Benno Mueller-Hill of Harvard isolated the repressor molecule that controls the lactose operon in *E. coli.* A year later Mark Ptashne, a Harvard colleague, found another repressor substance in a DNA virus called lambda, which lives in *E. coli.* Repressors turned out to be just about what Jacob and Monod predicted, proteins that combine with inducer molecules preventing them from binding to the operator. When this happens the operator is in an "on" position and the structural genes are making enzymes.

In 1974 Konrad Beyreuther and his associates in Cologne, West Germany, determined the exact sequence of the 347 amino acids in the lactose repressor protein molecule found in *E. coli.* Ptashne with the help of his colleagues completed the analysis of the lambda repressor with its 272 amino acids.

Research has also shown that there are specific sites in the operator gene to which repressors attach themselves. Six have been located in the lambda virus. It is now known that the site to which the repressor attaches itself determines which genes are turned off and which are turned on. Studies of gene regulation by repressors are receiving considerable attention.

PROMOTOR SITES

Until recently it was assumed that the transcription from DNA to messenger RNA was a simple, uncomplicated process requiring only one enzyme, RNA polymerase. It was thought that this enzyme moves along the length of one of the DNA threads and transcribes parts of it (genes) onto messenger RNA. We now know that to begin transcription, the enzyme must bind to a very specific part of the chromosome. This region, which is located next to the operator gene, is called the promotor.

SIGMA STARTERS

How does the enzyme "know" which gene to transcribe and where to begin its transcription? Some answers to the question of how RNA polymerase "recognizes" the gene to be transcribed were supplied in 1969 by Richard R. Burgess and Andrew A. Travers at Harvard, and John J. Dunn and Ekkehard K. F. Bautz at Rutgers University in New Jersey. The groups simultaneously discovered that a part of the RNA polymerase molecule is able to "recognize" the "start" signal in the promotor. It "instructs" the enzyme where to begin transcription. They also discovered that once the enzyme is properly placed, a part falls away from the molecule and the rest of the enzyme carries on transcription. The portion of the enzyme which recognizes and binds the entire enzyme to the promotor portion of the gene to be transcribed, they called the sigma factor. Released sigma factors are recycled and used by other RNA polymerases.

On the other hand, when a repressor molecule

binds to the operator, RNA polymerase next to it on the promotor is physically blocked and unable to move along the gene transcribing it. This explains how repressors turn off genes.

RHO STOPPERS

Sigma factors only aid in the initiation of transcription; they cannot indicate where the enzyme should complete making messenger RNA molecules. This prompted Jeffrey W. Roberts, another Harvard researcher, to look for a "finishing" factor similar to the sigma factor. Late in 1969 he announced that he had found such a protein molecule, which "tells" DNA polymerase that its work is completed, by releasing the enzyme and the RNA product. He named this the rho (pronounced *roe*) factor. Whether the rho factor works by recognizing the stop codons (UAA, UAG, and UGA) or some other terminal base sequence, and whether it operates alone or in combination with RNA polymerase is unknown. The molecular mechanism of rho has yet to be discerned.

SUMMARY OF GENE CONTROL IN MICROBES

The Jacob-Monod theory provided science with a working model of gene control in bacteria based upon feedback mechanisms. The model consists of a functioning unit (the operon) composed of three genes making lactose-digesting enzymes controlled by a fourth gene (the operator). A regulator gene, nearby, produces repressor substances which bind to the operator, holding it in an "off" position so that no mRNA for enzymes is being made. Inducer molecules, such as lactose, combine with repressor molecules, preventing them from binding to the operator and switching off. In this way the operator is switched on and the genes that make the lactose-digesting enzymes are functioning. In essence, lactose turns on the genes and lack of lactose turns them off.

The gene control theory has been verified in some ways and modified in others. It predicted the existence of messenger RNA and repressor substances, both subsequently discovered. However, the model is a vastly oversimplified version of the present-day concept of gene control. Specific repressor sites on the operator gene have been found and gene control by repressor substances is now being explored. A promotor gene next to the operator gene has also been discovered. Transcription by RNA polymerase starts at a specific area of the promotor. Repressors bound to the operator stop transcription by physically blocking the enzyme from moving along the gene to make mRNA. The transcribing enzyme is "told" where to begin transcribing by the sigma factor, which is part of RNA polymerase. The sigma factor recognizes and binds the enzyme to the promotor area of the correct gene and then drops out.

Another factor, rho, terminates the message. The mode of operation of the sigma and rho factors, and their relationship to the "start" and "stop" codons and to DNA and RNA polymerase, are still unclear and are areas of intense scientific investigation.

One must remember that the mechanisms for gene control are based upon research with simple organisms: viruses and bacteria. More complex organisms like ourselves probably have far more complex mechanisms involving hormones and nucleoproteins, to which we will now turn our attention.

GENETIC CONTROLS IN HIGHER ORGANISMS

The genetic material in simple organisms such as bacteria, viruses, and a few algae consists of minute threads of naked DNA immersed in the cytoplasm of the cell. In all other organisms, genes are packaged in chromosomes which are confined to the nucleus. These chromosomes are much larger, more numerous, and chemically more complex than those of microbes. They consist of bundles of DNA wrapped and bound to protein coverings. A chromosome contains about as much DNA as proteins and a very small amount of RNA. The chromosomal proteins are divided into histones and the nonhistones. These have been known for over 100 years and attracted very little attention.

HISTONES: GENE TURNER-OFF-ERS

In 1943 Edgar and Ellen Stedman at the University of Edinburgh, Scotland, suggested that the nuclear proteins called histones might be gene repressors. However, it was not until 1962 that the first scientific study of the effects of histones on gene activity was conducted. James Bonner and Ru-Chih Huang at the California Institute of Technology reported that experiments with the chromosomes of the garden pea plant showed that natural chromosomes containing histones make RNA in a test tube more slowly than chromosomes stripped of their proteins. Adding histones to naked chromosomes seriously reduced their RNA-making capacity.

In a parallel experiment Vincent G. Allfrey and Alfred E. Mirsky at the Rockefeller Institute in New York City got similar results with the chromosomes from the thymus gland of a calf. Stripped calf chromosomes yielded more test-tube RNA than chromosomes to which histones were added. These studies support the idea that histones are gene repressors; that they cover the genes like a blanket and therefore suppress RNA synthesis. It was also assumed that there were many kinds of histones, one to cover each kind of gene. The chemical findings contradict this last assumption. Only five groups of histones have been clearly identified.

Most astonishing is the similarity between the histones from completely different organisms. The amino acid sequences of the histones of higher plants and animals are almost identical. For example, only three small differences were found in the amino acid sequences of cow and pea plant histones. A 1977 study of the base sequence in the sea urchin by M. L. Birnstiel, W. Schaffner, and H. O. Smith at the University of Zurich, Switzerland, supports previous findings on the similarity of the amino acid sequence of histones among widely different organisms. In this case, the DNA coding for the amino acids in a sea urchin histone was very similar to that present in a calf. How the histone "blanket" is removed to allow a gene to act remains a puzzlement. Scientists scratch their heads and beards while they try to coax the answers to their questions out of histones. It seems likely that another kind of molecule may be implicated—nonhistones, the other chemical constituent of chromosomes.

NONHISTONES: GENE TURNER-ON-ERS

By definition, nonhistones found in the chromosomes are the proteins that are not histones. There are many more nonhistone groups than histones in a chromosome. Unlike the histones, which are almost identical among higher organisms, there are great differences in the kinds of nonhistone proteins present in the cells of different species as well as in different tissues within the same organisms. Experimental evidence supporting the idea that nonhistones play a part in controlling the activities of specific genes is accumulating. Changes in these proteins have been observed in cells where there are changes in gene activities such as growing and reproducing cells and in cells infected with cancer-causing viruses. Several studies have definitely shown that nonhistones stimulate transcription, that is, the production of mRNA from DNA. The most convincing evidence comes from reconstitution experiments in which the chromosomes are taken apart and the constituents are put together in various combinations. R. Stewart Gilmore and John Paul of the Beatson Institute for Cancer Research in Glasgow, Scotland, separated calf thymus chromosome into DNA, histones, and nonhistones. They then combined the DNA and the histones in a test tube. No transcription took place; no RNA was produced. When nonhistones were added, the chromosomes acted normally and produced calf thymus RNA.

They then obtained chromosomes from the bone marrow and the thymus gland of a rabbit. Each of the two kinds of chromosomes were broken down into the three basic constituents, DNA, histones, and nonhistones. The DNA and the histones derived from the bone marrow and the thymus gland chromosome were then mixed together. When bone marrow nonhistones were added to the mixture, the chromosomes which formed produced the kind of RNA that behaved like bone marrow RNA. When thymus nonhistones were added to a sample of this mixture, the chromosomes which reconstituted themselves produced RNA that acted like thymus RNA. These results strongly support the notion that a specific gene is turned on by a specific nonhistone.

It appears that genes covered by histones are "turned off." Nonhistones seem to "turn on" specific genes which they may do either by removing the histone blanket or by preventing histones from covering the genes. How nonhistones interact with histones and DNA to control specific genes is a mystery yet to be solved.

Histones have also become the focus of recent studies as part of the nucleoprotein fibers that make up the chromosomes in cells with a nuclear membrane, those of higher organisms. The R. Kornberg "beads-on-a-string" model of chromatin proposes that the "beads" are repeating units consisting of two each of four types of histone associated with 200 base pairs of DNA. The string is visualized as a flexible naked chain of DNA. The basic idea is that a chromatin fiber is a flexible joined chain of repeating units. It is also believed that the highly folded condition of DNA in these chromosomes is the result of the interaction of histones. As this model is untied, a better understanding will emerge of the structure and function of chromosomes in higher organisms. (See diagram on page 64.)

HUFFING AND PUFFING DNA

Is it possible to see genes at work? Can you tell when genes are turned on or off? The answer seems to be yes.

Many insects, including flies, go through several distinct stages in their development. Flies lay eggs which hatch into larvae, the feeding or caterpillar stage. Each larva changes into a pupa, the resting or cocoon stage. The pupa turns into the adult, the reproducing stage. About a century ago, giant chromosomes were found in the salivary glands of fruit fly larva. These giant chromosomes, a hundred times as long as and ten times as thick as normal chromosomes, caught the eyes of the geneticists not only because of their colossal size but also because of their banding patterns. Each kind of chromosome can be identified by a characteristic sequence of vertical light and dark bands. The dark bands are thought to be DNA and the light ones, proteins. The number of dark bands is regarded as representative of the number of genes in a chromosome, a concept still being debated. Attempts to penetrate the inner structure of chromosomes have not been too successful.

However, puffed-out areas along these giant chromosomes have been observed for some time but escaped proper interpretation until 1952, when

Wolfgang Beermann and Ulrich Clever of the Max Planck Institute in Germany, and Clodowaldo Paven and Martha E. Breuer at the University of São Paulo in Brazil, working independently, proposed that the chromosomal puffs are "turned-on" genes. They found different puffing patterns on the chromosomes from different tissues. They observed differences in the same tissue at different stages in the development of an organism. The puffs are the same at the same stage in the same tissues. Later Beermann showed that the unpuffed sections of the chromosomes contain only DNA while the puffs contain both DNA and RNA.

Continued investigation supports the idea that puffs consist of open loops of DNA actively transcribing RNA. In a study of a small fly that makes a protein from which the larva spins threads for its pupa in the cocoon stage, four specific puffs that produce these silk-like threads were located on one of the giant chromosomes. A definite link between a specific protein and specific puffs was established.

Further evidence supporting the theory that puffs are active genes comes from experiments with hormones in insects. The development of the various stages in the life history of an insect are controlled by a series of hormones. Ecdysone (EK-deh-sohn) is the growth hormone required for their development from the larva or caterpillar stage to the pupa or cocoon stage. In the early caterpillar stage, fruit fly chromosomes show very few puffs. A few minutes after the injection of ecdysone, existing puffs disappear and new ones appear. During the

Chromosome puffs: Puffs in the banding pattern of the giant chromosomes in the salivary glands of the fruit fly larvae.

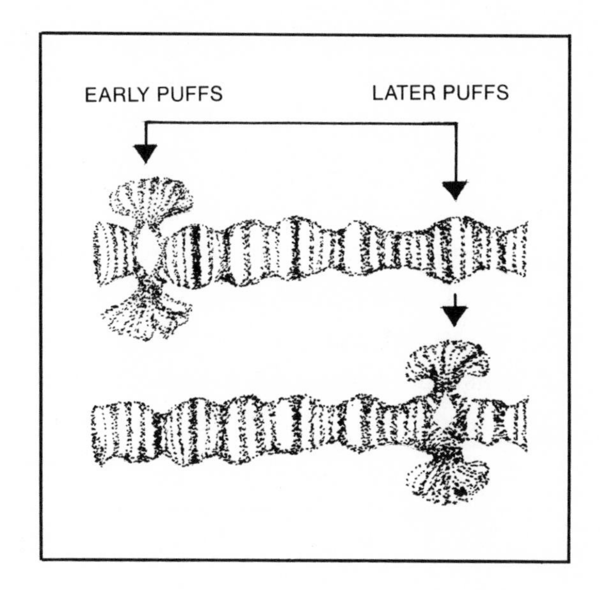

fly's development from the caterpillar to the cocoon stage over 100 puffs appear. Each chromosomal band puffs at an exact time and remains puffed for a precise interval. The puffing seems to be a direct response to the hormone. The huffing and puffing of chromosomes is still in the realm of scientific speculation with respect to how hormones produce their effects, but the story is hardly a nursery tale.

WILD AND WOOLLY CHROMOSOMES

Several years ago greatly elongated chromosomes covered with what appeared to be bushy fibers were found in the egg cells of amphibians. Because of their appearance they were called lampbrush chromosomes. The fibers are looped-out regions of double-threaded DNA. Extending from the loop are shorter fibers consisting of DNA and RNA. The fact that some loops are very distinctive in structure suggests that a loop may be a single gene coded to produce a particular protein. The fibers extending from the loops seem to be actively transcribing genes. Loops lacking these fibers are the inactive genes and those with increasingly longer fibers are increasingly longer RNA molecules attached to their DNA template. Since developing eggs are very active

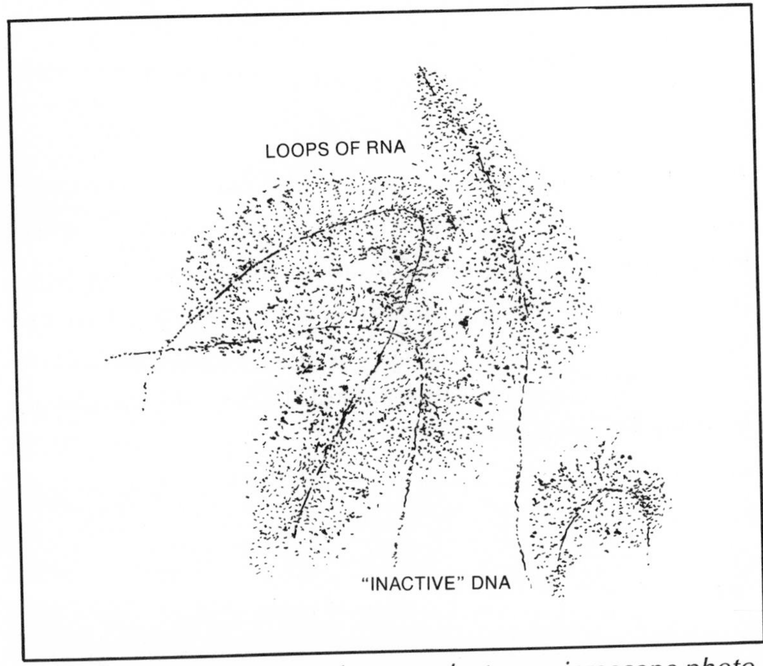

These are drawings taken from an electron microscope photograph of the lampbrush chromosomes in a frog egg. The long thread is DNA. The fibers extending from it are RNA molecules. Where no fibers extend from the DNA, it is assumed that no RNA is being made.

in protein production, groups of extended threads and side loops represent selected genes that are turned on.

The open loops which constitute the chromosomal puffs and the extended threads of the lampbrush chromosome are regarded as evidence of genes in action. In both instances the question being asked is "How are these genes turned on and off?"

SUMMARY OF GENE CONTROL IN HIGHER ORGANISMS

Little is known about gene control in higher organisms. What we do know comes mostly from research with viruses and bacteria. This research gives us some understanding of the cellular and molecular activities common to all life. Genetic control studies have focused on the chromosomes, which consist of DNA associated with nuclear proteins, the histones and nonhistones. The histones repress transcription while the nonhistones activate specific genes. The mechanism is obscure. The puffs and brushlike hairs which are associated with the chromosomes of a few insects are interpreted as DNA threads in the process of transcription. A growth hormone produced during the development of insects induces a succession of chromosomal puffs. We do not understand how this or any other hormone exercises gene control.

10 MISTAKEN DNA MESSAGES—MUTATIONS

There is, for the most part, a very strong resemblance between parents and their offspring. Baby zebras sport stripes just like their parents; young giraffes are as long-necked as their ancestors, and the hand of an infant is unmistakably human. However, every once in a while a "freak" appears among a group of organisms, dramatically different from its parents and the other members of the species. White blackbirds, six-toed kittens, seedless apples, and wingless flies are examples of these "oddballs." Sometimes these "freaks" have offspring that also are "freaks." Such inherited changes are called mutations (mew-TAY-shuns); individuals showing these unusual traits are referred to as mutants.

Mutations occur universally in all living things without exception. Normal five-fingered parents may have six-fingered children, and red bacteria may give rise to white ones. This tendency to be "different" is one of the characteristics of being alive; mutations are a built-in property of life.

MUTATIONS—A GENE CHANGE

Until the turn of the present century, there was no scientific explanation for these "freaks" which appear rarely but regularly in nature. Mutations seem to be mistakes or accidents which disappear, with others always cropping up to take their place.

One day, about 1910, a "freak" suddenly appeared among the fruit flies being raised and studied by Thomas H. Morgan and his workers at Columbia University. They were conducting experiments in heredity, using a tiny fly commonly found around ripe fruit in the summer. This "freak" had white eyes in sharp contrast with the red eyes of its parents and of all other fruit flies. By breeding the white-eyed fly with normal red-eyed ones, Morgan created a new variety of white-eyed fruit flies. What caused the mutation? Morgan examined the chromosomes of the mutants, hoping to find some differences between white- and red-eyed flies which would account for these mutations, but none could be detected. However, the mutations led to other studies and eventually to the modern chromosome theory of genetics, the idea that chromosomes consist of genes, units of heredity, arranged like beads on a string. In 1933 Morgan received the Nobel prize in medicine or physiology for his "discoveries relating to the hereditary function of the chromosomes." Mutations were explained as changes in the genes, but nothing was then known about their chemical or physical structure.

MUTATIONS—MIXED-UP DNA

Today the gene is described as a portion or segment along the DNA ladder consisting of a set sequence of bases coded to produce a particular protein. Mutations are now defined as the result of changes in the order of the bases in the DNA molecules. DNA makes endless replicas of itself without error, like a printing press turning out copy. It duplicates the exact sequence of the thousands and millions of bases which represent the genetic code for thousands of traits. However, every once in a hundred thousand or million duplications, something goes wrong, and an error is made copying one or more of the bases. A base may be changed, or omitted, or put in the wrong place, or an extra one added.

The following examples help to show how changing, omitting, adding, or inverting a base in the DNA chain alters the message.

Suppose we start with a message that reads:

EATTHEHAMANDEGGNOW

Let us assume that the message is written in three-letter words which are read from left to right with no overlapping letters. Its meaning is clear:

EAT THE HAM AND EGG NOW

If the letter H is added at the beginning of the sentence we read:

HEA TTH EHA MAN DEG GNO W

The sentence is nonsense and tells you nothing. If the first letter E is omitted the sentence reads:

ATT HEH AMA NDE GGN OW

Again the message is garbled and meaningless.

In any event, the genetic message is changed and a mutation may show up. The base sequence in a normal five-fingered boy may read:

ACG TTG CAT

while the copy may read and produce a six-fingered boy:

ACG TTG GAT

A "copy error" was made in reproducing the C in the last triplet and a G appeared instead. This may be the only difference in a chain of thousands of bases; it may change the entire meaning of the message and produce a six-fingered mutant. Once the "typographical" error creeps into the text of the DNA story and appears in the reproductive cells, the mutation is passed on to the offspring. Mutations occur at random in nature. "Mistakes" are made anywhere along the DNA strand. No base or bases are exempt.

MUTATIONS GOOD AND BAD

Since the discovery of the white-eyed mutation, hundreds of others have been found among fruit flies—flies without wings, without eyes, with yellow bodies instead of gray, and with purple eyes instead of red. Of course, only one fly in a hundred thousand or a million is a mutant. This may seem to make mutations a rare event. However, if you raise enough flies, mutations inevitably appear. The rate at which mutants show up depends upon the rate of reproduction of the species. Mutations appear more commonly in fruit flies than in humans because these flies produce a brood of a few hundred offspring every other week.

A very rich source of mutations for scientific investigation is found in viruses; they reproduce at the rate of ten times per minute. Bacteria reproduce every 20 minutes.

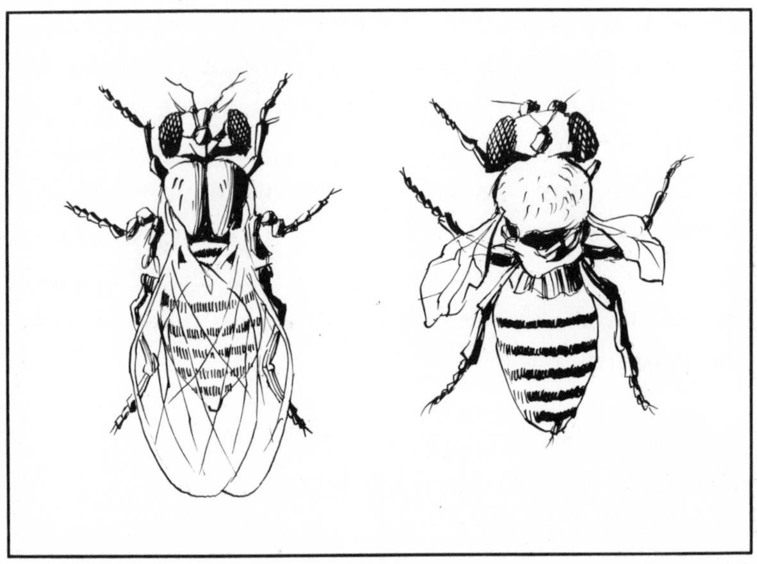

Left, *normal,* Right, *mutant fruit fly. One fruit fly in many hundred thousands is a mutant.*

Most mutations are "bad" for the organism; they put the mutant at a disadvantage in the struggle to survive. In nature, mutants rarely live long enough to reproduce; they are killed off by their enemies. A wingless fly would probably not live very long in competition with its winged competitors for food and a mate. White-eyed female fruit flies prefer to mate with red-eyed males rather than with white-eyed males.

On the other hand, mutation can also be "good," giving an organism an advantage over its competitor. When this happens, the mutation survives and is passed on to more and more individuals in the species. The mutation that makes a bacterium resistant to penicillin is good for that kind of bacterium.

Whether a mutation is "good" or "bad" depends upon how it affects the life of an organism. In a stable environment, mutations rarely improve the organism. In a changing environment, the right mutation at the right time may save the species from being wiped out. The penicillin-resistant strain of bacteria is an excellent example of this. This is "good" for the bacterium, of course, but "bad" for a human being made sick by this organism.

MUTATIONS AND EVOLUTION

From the very beginning of life about several billion or so years ago, mutations have been produced in an unending stream, and they are largely responsible for evolution. The millions of species of plants and animals which have appeared are the products of mutations which have accumulated in the DNA messages of these organisms and made them different. For the individual species, mutations are a kind of insurance against a rainy day. It may never have to use the mutations but there is no way of knowing which one of the thousands of mutants appearing at random can save the species from extinction.

MOLDS AND MUTATIONS

The ways that mutations affect an organism is most easily followed in simple living things, such as viruses, bacteria, and molds. Such work was pioneered in the 1940s by George W. Beadle and Edward L. Tatum of Stanford University. They used the common pink mold, neurospora (new-RAH-spo-ruh), which appears on bread in the summer. This mold can be grown in a test tube on a very simple diet consisting of a mixture of sulfate, phosphate, and nitrate salts, to which some table sugar and the vitamin biotin are added. From these few substances, neurospora can make amino acids, carbohydrates, proteins, fats, enzymes, and all the vitamins, except biotin—in short, everything it needs in order to grow and reproduce.

The molds were exposed to X rays to induce mutations and then put back on their simple diet. Some of them could no longer grow on this diet.

The X rays had altered some of the genes so that the mold could not make all the life-giving food substances. To find out exactly which substances the mutants could not synthesize, foods were added one by one until the mold began to grow.

For example, the X-rayed mold may have lost the power to make the vitamin niacin (NIGH-ah-sin), and therefore would not grow. When niacin was added to its diet, all was well again. Furthermore, when this disabled mold reproduced, all its offspring showed this same defect. The X-ray-induced mutation was inherited.

After performing hundreds of such experiments, Beadle and Tatum came to the conclusion that genes control the chemical reactions in a living cell through enzymes, which, as we know, are protein molecules. They believe there is a specific gene for the synthesis of each enzyme taking part in the chemical reactions in the body. In 1958, Beadle and Tatum were awarded shares of the Nobel prize in medicine or physiology for "their discoveries that genes act by regulating specific chemical reactions."

MUTATION MAKERS IN NATURE

What causes mutations? What is there in nature that constantly upsets the DNA molecules in all living things and induces mutations? Actually, we do not know the causes of most of the natural mutations although we suspect many things around us. We know that the rate of mutations can be increased substantially by exposing reproductive cells to certain influences such as heat, chemicals, and radiation.

Everything on earth is constantly exposed to certain natural radiation: the cosmic rays from outer space, the ultraviolet rays of the sun, radioactive deposits, and radioactive fallout from A-bomb testing. Cosmic rays, which constantly bombard the earth, are the most powerful and most penetrating particles of energy yet discovered. Every minute about eight of these rays pass through each square inch of your body. As these particles shoot through a cell in your body, they may hit a DNA molecule and knock off one or two atoms, thus changing the chemical composition of a base and altering the genetic message. If the target happens to be a sperm or an egg cell, the mutation will appear in the next generation.

Natural deposits of uranium and radium are distributed all over the world. The radiations given off by these radioactive minerals are capable of altering the DNA structure of an organism and producing mutations.

The atmosphere is full of radioactive particles dating back to the first atomic explosion in 1945. Most of these particles are quickly carried down to the earth by the rain and snow, while others remain suspended in air for years. As these radioactive particles rain down on us, they contaminate the air, soil, and water. Plants and animals absorb them, and they enter your body in the food you eat, the water you drink, and the air you breathe. Once inside your body, some radioactive atoms continue giving off radiations for years and may cause mutations. Strontium 90, one of the radioactive fallout atoms found in many foods and in water, tends to accumulate in the bones and emits radiations for years. Its half-life is 28 years; that is, it takes 28 years for strontium 90 to lose half its radioactivity.

MULLER'S MUTATIONS

It is no longer necessary to wait for nature to produce mutations. In 1927 Hermann J. Muller, then at the University of Texas, discovered how to make mutations by exposing fruit flies to X rays. He was able to increase the natural mutation rate at least a hundredfold. The kinds of mutations he produced included some never seen before, while others were the same as those which appear naturally. For example, Muller found some white-eyed flies among his mutants; they looked and behaved exactly like the natural mutants. In 1946, Muller was awarded the Nobel prize in medicine or physiology for his "discovery of the production of mutations by means of X-ray irradiations."

Muller could not make mutations "to order" or predict what kind of mutations his X rays would produce. He was shooting in the dark. There is no way of aiming the X rays at a particular DNA molecule and hitting a particular gene. As the X rays pass through the cell, they fan out and strike anything in their path. Some, by chance, hit a chromosome and knock out or rearrange a few atoms in its DNA chain, thus changing the genetic message and producing a mutation.

CHEMICALLY MADE MUTATIONS

Many chemicals are mutation makers. One chemical which can alter one of the bases in DNA and RNA, changing it into another nucleic acid base, is nitrous acid (HNO_2), which transforms cytosine (C) into uracil (U). This provides a method for making mutations "to order." Fraenkel-Conrat treated naked RNA strands of the tobacco mosaic virus (TMV) with nitrous acid and rubbed them into tobacco leaves. A whole host of mutants of TMV appeared. When the protein coat assembled by the RNA of one of these mutants was compared with that of the parent virus, 3 of the 158 amino acids making up the protein coats were different. Apparently the nitrous acid, by changing a C to a U, had altered the triplet code word for some of the amino acids in the protein chain. Let us assume that CCA is the code word for the amino acid proline and that nitrous acid changed it to read UUC, which is the code word for another amino acid, leucine. We can then explain why a substitution of proline by leucine was one of the three replacements in the mutant protein coat.

HUMAN MUTATIONS

Human beings are not exempt from mutations. Some of the ills of humanity are the result of disturbed and disarranged DNA molecules. Hereditary diseases such as hemophilia and color blindness are well known. Persons suffering from hemophilia are born with blood that clots with difficulty, if at all. Their chances of growing up to maturity are pretty slim, since they can bleed to death from a cut or a scratch. Color-blind individuals are normal in every respect except one—they cannot tell the difference between red and green. Both these conditions are due to mutations.

Since the 1940s we have discovered hundreds of human diseases that have a genetic basis. Some are extremely rare but others are far more common than suspected.

ENVIRONMENTAL POLLUTION AND MUTATIONS

No discussion of mutations is complete today without considering their major source, environmental pollutants. These genetic "change agents" are the products and by-products of our advancing industrial economy. They are polluting the air we breathe, the water we drink, and the soil whose products we eat. An increasing number of serious human diseases, including birth defects, genetic diseases, and cancer, have been traced to environmental pollutants. The fact is, more progress has been made in the conquest of germ than gene diseases. Why is this? There is no simple answer except to point out that more people are being exposed to more new kinds of pollutants that are both mutation makers (mutagens) and cancer causers (carcinogens: car-SIN-o-jinz).

Thousands of pollutants have been identified as actual or potential mutagens and carcinogens, and there are at least an equal number of new, untested polluting agents. Poisoning pollutants have been traced to radiations, industrial products and by-products, fumes from burning fossil fuels (natural gas, coal, crude oil, and gasoline), food additives and substitutes, pharmaceutical drugs, cosmetics, dyes, pesticides, and fertilizers. Prominent on the list of substances hazardous to human health are X rays, smog, smoke, soot, flame retardants, tobacco smoke, hair dyes, asbestos, Mirex, Red Dye II, PPB, DDT, PCB, mercury, and lead. Hardly a day passes without another suspect being added to the list.

There is open conflict and loud disagreement among people about how harmful these substances are to humans and other living things. In part this controversy arises from the kinds of "proof" presented by opposing groups. Evidence, for example, that smoking causes lung cancer in humans comes from controlled experiments with laboratory animals and from surveys of the incidence of human lung cancer among smokers and nonsmokers. Controlled experiments are conducted with nonhuman subjects; therefore, one may question how valid the results are for humans. Surveys involve humans but there are so many possible factors that may be related to cancer—age, sex, heredity, occupation, food, drugs, and so on and so on—that it is difficult to establish a cause-effect relationship. The proof is often described as circumstantial evidence or "guilt by association." Other factors feeding the fire of controversy are the vested interests of opposing groups, which may be personal, political, or economic. Despite these obstacles, "clean air," "clean water," and "pure food and drugs" are phrases emerging as goals if our planet is to survive.

MAN-MADE RADIATIONS AND MUTATIONS

The explosion of the first atomic bomb in 1945 set off a chain reaction that is still going on. It signaled the start of an international arms race of making and testing atomic weapons on the one hand, and of building atomic energy plants on the other—a sort of modern version of *War and Peace*. The result has been the creation and production of radioactive forms of every known element and the concentration of huge amounts of these materials. Worldwide contamination by radioactive fallout, although sharply reduced as the result of the test-ban treaty, is constantly being renewed by atomic testing programs and atomic missiles. Global "hot" deadly dust atoms require international monitoring of the amount and kinds of potential mutagens and carcinogens. Some radioactive particles are extremely shortlived, lasting only a fraction of a second, while others remain "hot" for hundreds or thousands of years.

Evidence of the damaging effects of radiation from radioactive substances comes from people working with or exposed to such high-intensity radiation sources. Survivors of the A-bomb attack, workers in atomic energy plants, X-ray technicians and radiologists, uranium mine workers, painters of radium dials on watches, and patients receiving large doses of X rays or radioactive therapy, are among those who have shown relatively high rates of blood cancer (leukemia) and bone and skin cancer. As instruments of war, atomic weapons are doubled-edged swords. They do not discriminate between friend and foe. Atomic weapons not only kill immediately but destroy and disable their victims indiscriminately for generation after generation.

The deadly effects and long life of radioactive materials have given us second thoughts about putting all our "eggs" into atomic energy plants as the ultimate solution to the growing energy crisis. This second look is prompted by a great unsolved technological problem, the disposal of the hot, long-lived "ashes" of those atomic energy "furnaces." For years radioactive wastes were dumped into the ocean or consigned to abandoned salt mines. But ashes are accumulating faster than they can be swept under the rug. The situation is serious enough to prompt the suggestion that perhaps radioactive wastes should be shot out into space where time and the rest of the universe will take care of our unwanted "garbage." Until we find out how to cool or recycle radioactive wastes, and learn how to control environmental hazards connected with such wastes, continued expansion of atomic energy plants must be carefully weighed.

No source of potentially harmful radiations is being overlooked. The long-time effects of dosage with X rays and ultraviolet rays, both of which are mutagenic, are being reevaluated to reduce hazards of overexposure. Emissions from television sets, microwave ovens, and high-tension electrical transmission wires are not beyond suspicion and are being studied as possible sources of mutagenic rays.

CHEMICAL CONTAMINANTS AND MUTATIONS

Probably the greatest sources of mutagens are the thousands of chemical compounds contaminating every part of the earth and all its inhabitants. There is considerable evidence that a large proportion of human cancer may be linked to toxic chemicals contaminating the environment, many of them yet to be tested for mutagenic and carcinogenic properties. This frightening situation is attributable to the fact that it takes years of laboratory testing and more years of field testing before all the effects and side effects of a chemical compound are known.

A SCREENING TEST FOR MUTATIONS

A promising procedure for detecting the mutagenic properties of chemical contaminants was devised by Bruce N. Ames of the University of California at Berkeley in 1975. This is a rapid, inexpensive screening technique that tells whether or not a chemical compound can induce mutations in a special strain of bacteria, *Salmonella typhimurium,* a much-studied organism that causes a typhoid fever-like disease in mice. There is evidence from many sources that, with few exceptions, cancer-causing chemicals are mutagens. Hence a chemical that causes mutations in *Salmonella* is presumed to be carcinogenic until proven otherwise and is therefore screened out for further testing with laboratory animals.

Over 300 kinds of chemicals, cancer causers and non-cancer causers, have been tested for mutagenic attributes by the Ames technique. Over 90% of the chemicals known to be carcinogens from other tests proved to be mutagenic. Only a few of the 106 noncarcinogens tested were mutagenic. Mutagens with carcinogenic potential were found in cigarette smoke, hair dyes, and soot from city streets. In 1978 a group of scientists at the University of California at Davis, using the Ames method, reported the results of testing the mutagenic powers of fine ash dust obtained from a power plant in the West burning pulverized low-sulfur coal. They found that this fine coal ash induced mutations and hence represented a potential carcinogenic threat to human health, possibly causing lung cancer.

ENVIRONMENTAL POLLUTANTS AND CANCER

The close relationship between the mutagenic and carcinogenic properties of increasing numbers of chemicals contaminating the environment is being established. These studies are providing clues about environmental causes of cancer in humans and other organisms and ways and means of controlling this affliction.

The World Health Organization estimates that 80% to 90% of all human cancers are related to or induced by environmental agents. A very close relationship has been found between the incidence of cancer and the degree of industrialization: the greater the concentration of industry, the higher the incidence of cancer in that area. Robert Hoover and Joseph Fraumeni, Jr., of the National Cancer Institute reported the following findings in a 1975 survey of occupational cancer:

1. High rates of bladder cancer occurred in the vicinity of factories making hair dyes, pigments, drugs, perfumes, cosmetics, and toiletries.
2. Lung cancer was high in the areas manufacturing industrial gases, medical drugs, soaps and detergents, paints, pigments, and synthetic rubber.
3. Liver cancer was high among persons living near or working in factories making synthetic rubber, soaps and detergents, cosmetics, printing ink, and certain organic compounds.

More than 1500 known or suspected carcinogens have been identified in American plants and factories; of these, only 7 are currently regulated. A 1976 survey by the National Institute for Occupational Safety and Health reports that 1 million American workers may be exposed to known carcinogens in the factories where they work. Another 20 million—one out of four—are exposed to potentially dangerous pollutants. The survey also revealed that the most dangerous industries in this country may be those that manufacture scientific and industrial equipment where workers are exposed to solder, asbestos, and thallium (a poison). The latter is considered one of the most toxic chemical elements. Interestingly, the chemical industry ranked 12th on this list.

Lung cancer is the most rapidly increasing type of cancer in this country and the leading type of cancer in American males. About 100,000 Americans will develop lung cancer this year and 90% of them will die within 3 years; most of them are cigarette smokers.

Another related finding is a report by the Louisiana State University Medical School in New Orleans. In a stretch of the Mississippi River which provides drinking water for 1½ million people they found that the water contains chemicals that "may be present in sufficient quantities to induce cancer." The greatest concentration of chemical pollutants was in the waters around heavily industrialized areas. The state with the highest concentration of industry in this country has the distinction of also having the highest rate of cancer.

ESTABLISHING THE CHEMICAL-CANCER CONNECTION

Several difficulties confront investigators in attempting to establish the link between chemical contaminants and cancer. Not only are years of laboratory and field testing required, but sometimes harmful effects surface only after decades have elapsed. Asbestos, which is linked to cancer and other diseases in humans, may not produce ill effects among asbestos workers until 20 or 30 years after exposure. DDT when it first appeared over 30 years ago was hailed as a miracle drug because it was so effective against insect carriers of typhus fever and malaria and insect pests that destroy our food crops. Today DDT is banned in the U.S. for almost all uses: it is known to accumulate in animal tissues where it is very toxic, and it is also a suspected carcinogen.

The DES story is equally distressing. DES stands for diethylstilbestrol, a synthetic female sex hormone widely used in the 1950s to fatten cattle, sheep, pigs, and poultry for human consumption. When it was found that small amounts of DES caused cancer in laboratory animals, it was banned as an animal food in 1973. Under the Delaney clause, an amendment in 1958 to the Food, Drug, and Cosmetic Act, any carcinogenic food additive regardless of the amount that causes cancer in man or in any laboratory animal cannot be sold for general use in foods or beverages. Natural carcinogens are exempt. The Delaney clause, limited as it is to carcinogen and to food additives, is criticized by some as too restricted in its application. For example, saccharin banned under the Delaney clause will only get a warning label.

To return to the DES story, it has been discovered that a form of cancer of the human female reproductive tract appears in some young women whose mothers were treated with DES during their pregnancy to prevent miscarriages. These carcinogenic effects have not appeared in the DES-treated mothers but in the "DES daughters" 20 years later. Several studies since 1971 have connected DES with increased cancer of the reproductive system in daughters, possible birth defects in sons, and an increased occurrence of breast cancer in the mothers themselves.

Another problem is chemical interaction between chemical contaminants. For example cyanides, which are found in industrial wastes, are toxic to water plants. In the presence of zinc or cadmium, the toxic properties of cyanides are intensified.

Some chemicals are in the environment in such small quantities that they would seem to have no harmful effects. However, if they have a tendency to accumulate in the body of organisms the accumulation may reach deadly levels, for example with lead or DDT. Other chemicals are neither mutagens nor carcinogens outside the body, but when they enter a cell are chemically changed into molecules that are mutagens and powerful carcinogens. For example, nitrates cause no genetic change; but in human cells they are converted to some extent into powerful carginogens (nitrosamines) capable of altering the DNA base guanine (G). There is little doubt that many carcinogens cause cancer by changing the DNA in a cell.

SUMMARY

Mutations are the random result of changes in the bases or their order in the DNA (or RNA) molecule. Mutant offspring may exhibit traits strikingly different from those present in the parents. This phenomenon takes place naturally in all living things and is the basis of evolution. Most mutations occur "spontaneously"—that is, their causes are unknown. However, mutations can be triggered by several agents in the natural environment: cosmic and ultraviolet rays, solar radiations, deposits of radioactive mineral such as uranium, and chemicals.

A half century ago, Muller discovered that the rate of mutations could be increased in fruit flies by exposing them to X rays. Some artificially induced mutants were identical with those found naturally. Other mutation-causing agents have been discovered, including chemicals capable of altering DNA bases.

Attention is being focused on environmental contaminants with mutagenic powers—radiation and chemicals. These contaminants are largely the products and by-products of modern technology; they have been linked with such serious human illnesses as birth defects, genetic diseases, and cancer. Establishing a cause-effect relationship between a contaminant and a disease or disability, for example, cigarette smoke and lung cancer, is difficult. Evidence comes from experiments with laboratory animals and from surveys of humans; both methods yield circumstantial evidence which is open to questions and interpretations.

Radiation from atomic energy sources—atomic weapons and atomic energy plants—is a powerful mutagen and carcinogen. Persons exposed to high-energy radiation sources show relatively high rates of cancer. Radioactive fallout from exploded atomic weapons has affected the entire planet and requires continued international monitoring and controls. The "spent" fuels from atomic energy plants remain "hot" for long periods of time. Their disposal remains an unsolved problem and an obstacle in planning the use of atomic energy to solve the growing energy crisis.

Chemical contaminants are probably the most important source of mutagens with carcinogenic potential. Using a recently developed method for testing the mutagenic properties of chemicals employing special strains of bacteria, it was found that 90% of known carcinogenic chemicals were mutagenic, including cigarette smoke, hair dyes, soot, and fine coal ash.

A high positive correlation has been established between the degree of industrialization and the incidence of cancer in the surrounding neighborhoods. More than 1500 known or suspected carcinogens have been identified in American industrial plants—only 7 are regulated. Lung cancer is the most rapidly increasing type of cancer and is the leading kind of cancer afflicting the American male. Most victims are cigarette smokers.

In the linkage of chemical contaminants to human disease there is often a long time interval between exposure and illness. Chemicals such as the insecticide DDT, the synthetic sex hormone DES, and the mineral asbestos may show their effects only after decades have elapsed. Other chemicals interact in the environment and thereby increase their toxicity. Still others accumulate in the body tissues and become increasingly toxic and carcinogenic. And there are chemicals like nitrates which are harmless until they enter the body and are converted by body chemistry into powerful carcinogens.

11

DNA AND DISEASE

In some of the malaria-ridden areas of central Africa, almost half of the natives suffer from a curious hereditary blood disease. The red cells in their blood stream have a normal disk shape in the lungs and arteries where oxygen is plentiful, but as soon as they pass into the capillaries and veins, which contain less oxygen, the cells collapse. Many of these red blood cells assume the shape of a curved knife blade or sickle and are therefore called sickle cells. The body destroys these defective and deformed sickle cells, which seriously reduces the number of circulating red cells, causing a severe anemia. This disease, sickle-cell anemia (SCA), is often fatal. It is caused by the mutation of a single gene. Persons suffering from SCA have two genes for sickling (SS). Those with one gene for sickling and one for normal (SN) show no symptoms of the disease—they are "carriers" whose descendants may be SCA victims. Those with two genes for normal (NN) have no sickle-cell problem.

MISSHAPEN MOLECULES

SCA has attracted considerable scientific attention because it is a "molecular disease" in which the

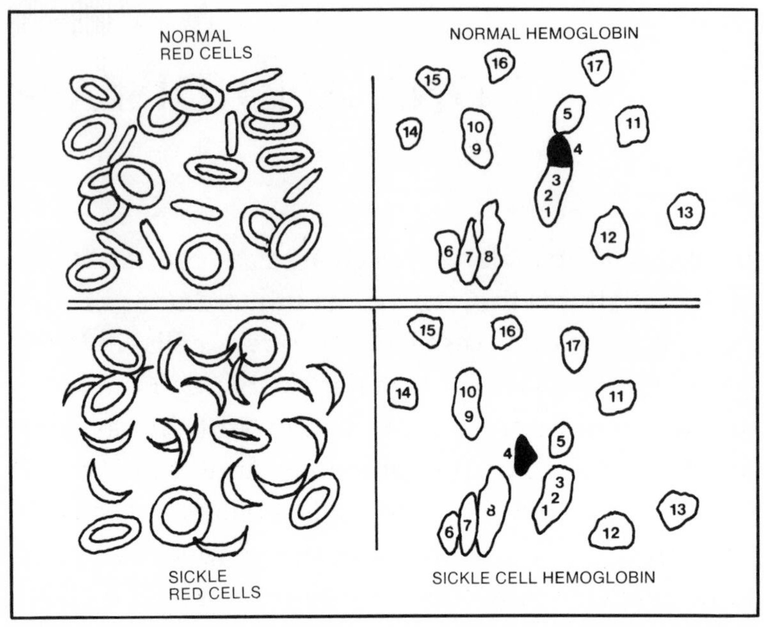

Normal and sickle cell: "Fingerprints" of normal and sickle cell hemoglobin differ only on one spot, that is, in one amino acid.

mutation of a single gene alters the structure of a complex protein, hemoglobin, the oxygen-carrying substance in red cells. In 1958 Vernon M. Ingram and his associates at Cambridge University, Cambridge, England, discovered the single change in the protein chain that completely alters the shape of the hemoglobin molecule. They compared the chemical makeup of the hemoglobin from a sickle cell with that of a normal red cell. First, by means of enzymes, they cut the long coils of sickle-cell hemoglobin into 26 smaller pieces. Then they placed a drop of solution containing these fragments on a moist piece of filter paper and passed an electric current through it. The pieces moved away from one another and spread out in a line across the paper. This sheet of paper was placed in another solution so that the edge with the spread-out pieces was moistened. The solution crept up the paper, carrying the fragments with it.

The hemoglobin pieces ascended the paper, but at different rates, and came to rest at various levels. By spraying the paper with a dye, the spots became visible and a pattern of smear spots resembling fingerprints appeared.

Each kind of protein has a different "fingerprint" by which it can be recognized and identified. Hemoglobin from a normal cell was "fingerprinted" and compared with that of sickle hemoglobin. The fingerprints were the same, spot for spot, except in one place. The two mismatched spots were then analyzed and each was found to contain eight amino acids, seven of which were the same. Again, there was just one difference: the sixth amino acid in normal hemoglobin was glutamic acid and the sixth place in the sickle chain contained valine.

Normal: Val-His-Leu-Thr-Pro-GLUTAMIC ACID-Glu-Lys
Sickle: Val-His-Leu-Thr-Pro- VALINE -Glu-Lys

The difference in one amino acid in a chain of 300 makes the difference between health and disease, between life and death. Just one wrong amino acid link in the protein chain of life and it collapses. Here is evidence that the order of the bases in a particular gene determines the order of the amino acids in a protein molecule. The Crick-Watson model of DNA and the concept that the order of the bases fixes the sequence of the amino acids in a protein explain the difference.

MISSING MOLECULES

Many of the chemical reactions in your body take place step by step in assembly line fashion. For example, the color of your skin, hair, and eyes is due to a brownish-black coloring material called melanin (MEL-a-nin). Dark-skinned people manufacture an abundance of this pigment; those with blond hair, blue eyes, and a very light complexion make much less. However, all people make some melanin. This pigment is produced from the amino acid tyrosine (TIE-row-seen), which is made from still another amino acid, phenylalanine. There are at least two chemical steps in producing melanin. Phenylalanine is changed to tyrosine, and tyrosine is converted to melanin, and at least two enzymes are needed, one for each step. If either enzyme is missing, the chemical production line is halted and melanin is not produced. Occasionally children are born with white hair, very fair skin, and pink eyes. They are referred to as albinos (al-BUY-noze). Their bodies lack the ability to make the enzyme necessary for converting tyrosine to melanin. Hence they lack coloring material in their hair, skin, and eyes (the eyes are pink because there is no pigment in them and what you see is the color of the blood in the back of the eye). Albinism is a hereditary condition resulting from the mutation of a single gene. Albino children may come from parents with black, brown, or normal white skins. This hereditary condition, as the result of a mutation, is a striking example of how the loss of a single enzyme among thousands can produce startling results.

INBORN ERROR IN METABOLISM—PKU

The mutation of a single gene is also responsible for one type of feeblemindedness caused by the inability of the body to metabolize the amino acid phenylalanine. Because the enzyme which normally oxidizes phenylalanine is missing at birth, this amino acid accumulates in the body and is excreted in the urine. The condition is known as phenylketonuria (fen-il-keet-a-NYUR-ee-uh), or PKU. The mutation deprives the organism of its ability to synthesize the missing enzyme, without which body chemistry is seriously disturbed, leading to brain damage.

Fortunately, PKU can be detected in infants a few weeks after birth and before phenylalanine affects the brain. By feeding PKU infants a diet low in phenylalanine, all the physical and mental symptoms are suppressed. The "inborn error in metabolism" can be corrected and these children can live normal lives.

More than 3000 human hereditary diseases have been identified. Victor McKusik of Johns Hopkins estimates that over 100 are the result of a single-genetic defect.

DNA and RNA are suspected of having some relation to cancer, a group of diseases in which some of the cells in an organism suddenly go berserk and begin a career of uncontrolled reproduction. Normally, cells are law-abiding members of the body community, subject to the rules and regulations set down by the organism. Each cell has its place and its responsibility written down in its DNA code. Skin cells have a definite size, shape, function, and rate of reproduction. New skin cells are produced only where and when they are needed by the organism. However, a skin cell may suddenly embark on a reproduction rampage in utter disregard for the other skin cells around it. It may produce cells completely unlike its neighbors. These cells are outlaws. They create other cells which follow in their footsteps. Within a short time, a whole mass of wildly dividing cells may pile up on the skin to form a lump of cancerous tissue, a tumor. One or more of these rabid reproducers at the bottom of the cancerous tumor may break away, anchor itself elsewhere in the body, and continue its cancerous career of corrupting and disrupting the living organism. The unlimited capacity of cancer cells to reproduce eventually leads to the death of the host and their own death. Cancer is not a single disease but many, which share a common characteristic of abnormal cellular growth. Almost 100 different kinds of cancer have been identified in humans.

CANCER-INDUCING DNA

What makes a normal cell turn into a raving, reproducing cancer cell? There are so many suspects that you can list them alphabetically, starting with "A" for atomic fallout, "B" for bruises, "C" for cigarettes, and so forth down on to the end of the alphabet with "Z" for zirconium. Regardless of who the guilty party or parties are, cancer is triggered by a change in the genetic instructions of a cell. A cancer cell is distinctly different from a normal cell in structure and function. In addition, cancer cells breed more cancer cells like themselves. The altered genetic code which spells out cancer is duplicated and passed on to all the daughter cells.

Cancer has been induced experimentally in plants and animals by a whole host of influences, among which X rays and chemicals are the best known. Many of these cancer-causing agents change the genes and chromosomes of the cell. Heavy doses of cancer X rays, for example, cripple the chromosomes and in some instances create cancer cells.

VIRUSES AND CANCER

In addition to physical and chemical agents there are biological agents associated with cancer, namely, viruses. The idea that cancer is an infectious disease caused by a virus has been advanced from time to time. The popularity of the virus theory of cancer has waned and waxed like the phases of the moon. It was not until 1911 that it was shown that a virus caused a kind of cancer, in chickens. Since then other viruses have been discovered that produce cancer in rats, mice, frogs, deer, rabbits, hamsters, monkeys, cats, dogs, horses, and cows—in practically all animals except humans.

How do viruses cause cancer? We do not know but we can make some educated guesses based on what we know about viruses, genes, and the cell. Both viruses and genes are bits of nucleoproteins with the genetic instructions contained in the nucleic acid. A virus injects only its nucleic acid into a cell; once its DNA (or RNA) is inside the cell any one of several events may occur. The viral DNA may take over and use the cell to make hundreds of other viruses like itself, in which case the cell is destroyed. Or the viruses may cause one of several diseases: common cold, influenza, chicken pox, measles, or poliomyelitis. Or a virus may attach itself to bacterial DNA and "pretend" to be part of the genetic apparatus of its host. Viral DNA can live in peace and harmony within the bacterium for many generations without making its presence felt or known. However, exposing such virus-carrying cells to X rays or chemicals sometimes "awakens" the concealed particles and they suddenly begin to reproduce and destroy the cell. Finally a virus may make a cell lose its control and plunge it into a wild orgy of reproduction forming strange and abnormal cancer cells. Possibly this is how viruses in all animals including man launch a cell on a cancerous career.

SUMMARY

Disorders in the DNA molecule are responsible for genetic disorders and diseases. Over 3000 genetic disorders have been identified in humans and of these almost 100 are due to changes in one single gene or another. Among the best-known human genetic diseases is sickle-cell anemia, a hereditary blood disease caused by a change in a single gene. Albinism, the lack of skin pigment, is another hereditary condition resulting from single-gene mutation. The gene affected directs the synthesis of an enzyme that makes melanin, the pigment found in skin, hair, and eyes. Still another single-gene mutation causes PKU, a condition in which the body is unable to metabolize phenylalanine, an amino acid.

Cancer is a cellular disorder in which there is unrestricted and abnormal growth of cells triggered by DNA disorders. There are about 100 kinds of human cancer. The three main sources of cancer-causing agents (carcinogens) are chemicals, radiation, and viruses.

Viruses have been found associated with cancer in practically all animals except man. Viruses appear to induce cancer by altering the genetic message in DNA. Although the human cancer-virus connection has not been established, so much important research has been done with viruses that they warrant further consideration. Hence the chapter that follows discusses some aspects of DNA and RNA cancer-causing viruses.

DNA AND RNA CANCER-CAUSING VIRUSES

Today, viruses are accepted as one of the three main cancer-causing agents, the other two being radiation and chemicals. As more and more cancer-causing viruses are tracked down in more and more animals, the virus-human cancer connection seems inescapable. Despite the best efforts of science, human cancer viruses defy detection. Once these viruses enter the cell, they disappear by shedding their protein coat and become just another part of the DNA living in peaceful coexistence with their host. There is no agreement among the experts on viruses as the cause of some forms of human cancer. Opinions and theories range from naming viruses as the culprits that cause all human cancer to seeing them as innocent bystanders that merely trigger cells with the potential to "go" cancer.

The search for the missing links between "vanishing" viruses and human cancer has led to many unexpected revelations and exciting discoveries which may lead to "breakthroughs" in the conquest of this dread disease.

There are two broad categories of cancer-causing viruses, DNA viruses and RNA viruses. They differ not only in the kind of nucleic acid they carry but also in their effects upon their victims. One group of DNA viruses gives rise to wartlike growths on many kinds of animals from mice to men. Herpes (HER-peas), another kind of DNA virus, is responsible for cold sores, fever blisters, and is associated with the kissing disease (mononucleosis), a rare jaw tumor in African children, and other types of human cancer. The best known of the DNA viruses are mouse and monkey viruses. The mouse virus polyoma (polly-OH-mah), is so named because it is capable of producing many kinds of cancerous growths in rats, mice, and hamsters. The monkey virus SV40 causes a great variety of tumors when injected into newly born laboratory animals.

THE MONKEY VIRUS—SV40

SV40 is a simple virus, spherical in shape, about 450Å or one 500-thousandth of an inch in diameter. It is covered by a warty protein coat that makes it look like a mulberry. Enclosed is a single circular double-stranded DNA chromosome coded for three genes. It has the barest necessities of life, and is among the smallest DNA-carrying viruses. When SV40 enters a monkey cell, it generally "disappears." Occasionally it begins to multiply, filling the host cell with millions of virus particles and destroying it. In mouse and hamster cells grown in tissue cultures, SV40 does not multiply—instead, it makes

Types of viruses.

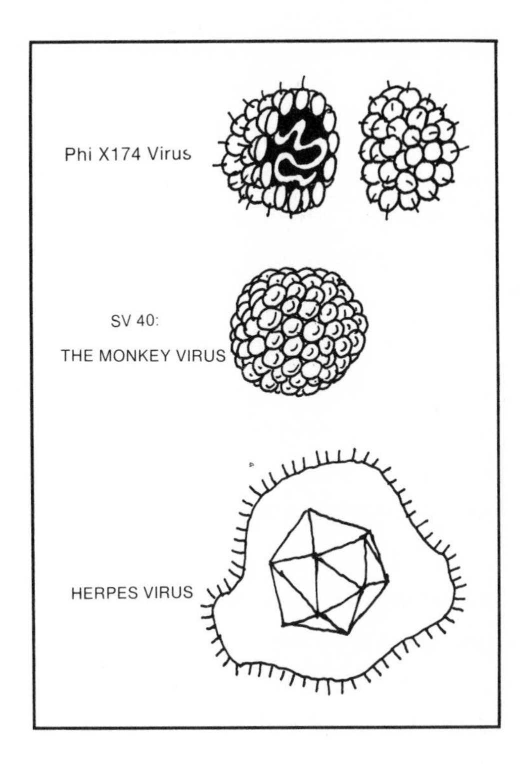

the cell cancerous. John Enders at Harvard University showed that human kidney cells not only support the growth of the SV40 in tissue cultures but become cancerous. Hilary Koprowski of the Wistar Institute inoculated cells from human donors with this virus and grew them in tissue cultures. Some of the cells turned cancerous. Some of the cancerous cells were injected under the skin of the donors. Small nodules appeared on their skin and under the microscope one of the nodules looked cancerous. This is about as close as we have come in trying to produce cancer in man experimentally.

THE MOUSE VIRUS—POLYOMA

The polyoma virus closely resembles SV40. In 1959 Renato Dulbecco and Marguerite Vogt of the California Institute of Technology began studying this virus with newly developed tissue culture techniques. They observed that this virus behaves much like SV40; it kills most cells while producing new viruses in prodigious numbers and transforming a few of the surviving cells into cancer killers. Sarah Steward and Bernice Eddy at the National Cancer Institute, who discovered this virus, found that polyoma is one of the most versatile cancer-inducing viruses and a powerful producer of antibodies. You cannot always find polyoma particles, but you can follow the trail of antibodies that they leave behind. When these two investigators stripped this virus of its protein coat and injected naked DNA into laboratory animals, some contracted cancer. Polyoma DNA was the first molecule which proved able to both produce a virus infection and to induce cancer, an observation that made many investigators take it seriously.

COLD-SORE VIRUS

The herpes viruses are linked to specific forms of cancer. They are relatively large as viruses go, about 1500Å, or one 150-thousandth of an inch in diameter. An outer envelope encloses a 20-sided protein

coat which houses a single linear double-stranded DNA molecule containing enough bases to code for about 100 average-sized protein molecules. Its lifestyle is typical of DNA viruses: it kills most of the host cells and tranforms a few into cancer cells. Transformed cells contain fragments of viral DNA which become part of the host DNA. Once a person has a herpes infection, the viral DNA seems to remain indefinitely in a "sleeping" stage, but it can come to life when stimulated by an external factor such as excessive sunlight. This may explain why people who get too much sunshine are prone to develop cold or fever sores and, in some cases, skin cancer.

CHICKEN CANCER

In 1967 the first conclusive proof that a herpes virus causes cancer was reported. This virus produces a cancerous growth in chickens that kills them. Normal cells are usually changed into cancerous cells that do not produce virus particles. Curiously, the virus reproduces in normal skin cells that are shed. These dead, virus-filled skin cells are highly infectious and spread the disease. Chicken cancer has been completely wiped out by a vaccine containing weakened chicken viruses and a related virus from a turkey. This suggests the possibility of cancer vaccines for humans, a long-sought-after but unrealized goal of cancer researchers.

HUMAN JAW CANCER

Another link in the virus-human cancer connection is the striking similarity between the chicken cancer virus and a herpes virus associated with human jaw cancer. In 1958 Dennis Burkett, a missionary surgeon in Uganda, observed a large number of African children suffering from jaw cancer. The disease is limited to mosquito country in hot and wet regions—not in the deserts or mountains. Suspicion fell on a virus spread by an insect, similar to the virus of yellow fever, which is transmitted by a mosquito. Soon cases were reported from other parts of the world where mosquito-transmitted diseases are practically nonexistent, indicating that the disease is infectious but not necessarily spread by mosquitoes. The hunt was on to find the virus in the cancerous tissue. It could neither be seen, nor grown, nor used to infect animals. In 1968 M. A. Epstein and Y. M. Barr of the University of Bristol, England, succeeded in growing jaw cancer cells and cells that produce this virus. The culture-grown viruses, known as EBV (Epstein-Barr virus) is a prime suspect as a viral cancer agent.

THE KISSING DISEASE

In 1968 Werner and Gertrude Henle and V. Diehl of the University of Pennsylvania disclosed that the "kissing disease," mononucleosis, not only is

caused by a herpes virus but cannot be told apart from EBV. It may be that the virus that causes the kissing disease and EBV are one and the same. However, almost everybody is infected by EBV, including those who have never kissed and never developed cancer. EBV in and of itself does not seem to cause human cancer. Nevertheless, herpes is high on the list of "most wanted" viruses in the search for the criminal causing human cancer.

RNA VIRUSES AS CANCER VENDORS

RNA cancer-causing viruses attack many kinds of backboned animals—chickens, rats, mice, monkeys, even snakes. These viruses all look much alike. They are spherical in shape, about 1000Å or one 250-thousandth of an inch in diameter. There is an outer membrane covered with knoblike structures and an inner core consisting of a tightly coiled strand of RNA surrounded by a 20-sided protein shell.

The most famous and longest known among the RNA cancer causers is the Rous chicken cancer virus. It was discovered in 1911 by Francis Peyton Rous at the Rockefeller Institute. He was the first to demonstrate that viruses can cause cancer in an animal. His discovery remained an academic curiosity and attracted little attention for 30 years. Belated recognition for his work came in 1966, when at the age of 87 Rous shared the Nobel prize in medicine or physiology. About 1940, Ludwik Gross, working at

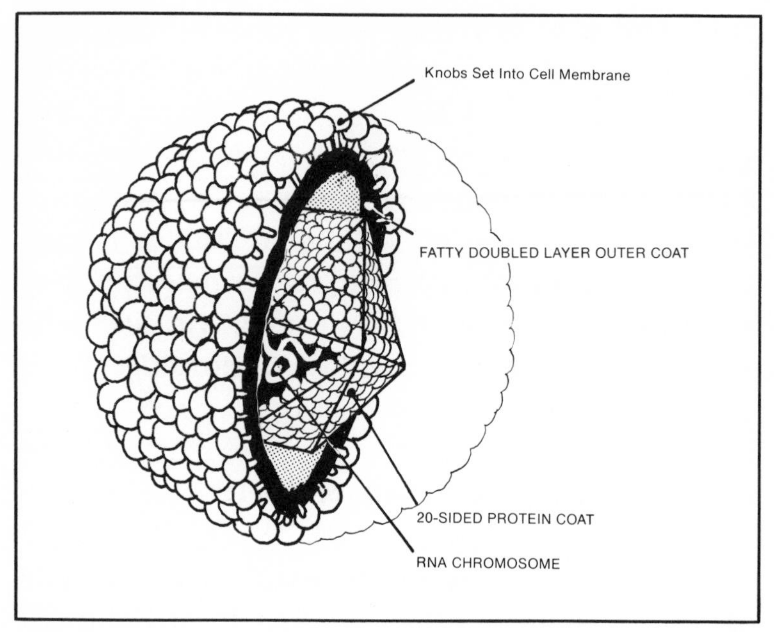

RNA virus.

the Veterans Administration Hospital in the Bronx, New York City, found a virus that causes cancer of the blood, leukemia (lew-KEY-me-ah), in mice. Since then, many cancer-causing viruses have been found in animals but none in humans.

A feature that distinguishes RNA from DNA cancer viruses is the ability to reproduce in the host cell without killing it or interfering with its normal functions. Perhaps the most startling discovery about RNA viruses is that they produce an enzyme that makes DNA from RNA. This finding runs contrary to the so-called "central dogma" of molecular biology, which holds that DNA makes RNA and not the other way around. In 1964 Howard M. Temin, at the

University of Wisconsin, suggested that the RNA in the Rous cancer virus was making strands of DNA using its own RNA as the template. This idea was regarded as biological heresy and was not well received by the scientific community.

In 1970 Temin and his coworker, Satoshi Mitzutani, did the "breakthrough" experiment and isolated the "RNA-to-DNA" enzyme. At the same time, David Baltimore at MIT announced the same results working with an RNA virus that causes leukemia in mice and with the Rous virus. The enzyme was named reverse transcriptase since it transcribed in the reverse direction, from RNA to DNA rather than from DNA to RNA. Additional support came from a third source, Sol Spiegelman of Columbia University, New York. He mixed the four DNA bases—A, C, T, and G—labeled with a radioactive tracer, viral RNA as the template, and the RNA-to-DNA enzyme. The product was labeled DNA for each of the six different RNA cancer viruses he investigated. This was conclusive evidence that the RNA in cancer-causing viruses can make DNA.

These studies opened up new avenues of cancer research. RNA viruses that make the RNA-to-DNA enzymes cause cancer in animals. For a time there was high hope that these RNA viruses were the key to understanding and controlling the cancer process. Such hopes were dashed by subsequent findings that the enzyme is present in noncancerous RNA viruses and that RNA viruses do not necessarily produce cancer. DNA copies of virus RNA have been

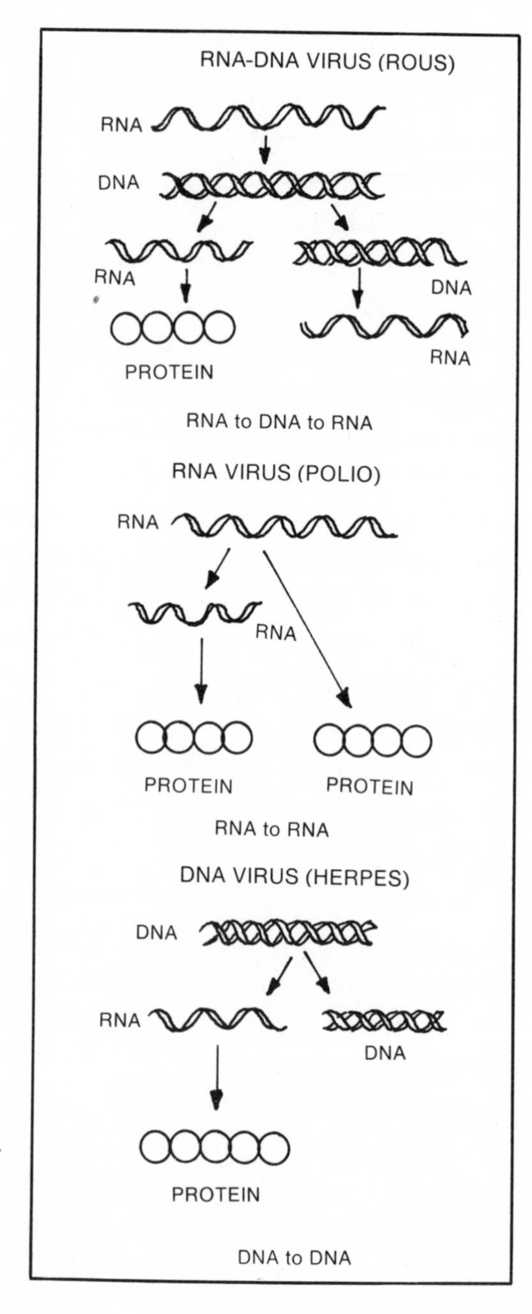

Three types of viruses: DNA, RNA, and RNA-DNA.

located in the DNA of host cells where they live quietly and do not make "cancer" waves. The fact remains that cells made cancerous by RNA viruses seem always to contain viral RNA. The RNA-to-DNA enzyme is necessary, if not sufficient, to cause cancer. This enzyme is implicated in the crime of causing cancer but how it commits its evil deed continues to elude investigators. The hunt for the vicious vanishing virus persists.

For their discoveries "of how cancer viruses and the genes of the host cell interact as possible links between viruses and human cancer," Renato Dulbecco, Howard Temin, and David Baltimore were awarded the 1975 Nobel prize in medicine or physiology. They showed that the DNA and RNA cancer viruses follow a common path at least part of the way to cancer. Their genes become part of the genes of the cells they transform into cancer cells.

13

DNA ENGINEERING— TEST-TUBE GENES AND VIRUSES

Can genes be isolated or made to order in a test tube or changed? Can living things be synthesized in a laboratory? Not too long ago these ideas were in the realm of science fiction or at best the "impossible" dreams of some scientists. Today all this and more has been achieved. Molecular biologists can pick whole genes out of a chromosome and use them as models for making synthetic genes that work. The complete genetic structure of at least two viruses has been determined, paving the way for making test-tube copies of them. The foundations of the science of genetic engineering are being built with the potential for altering and replacing the "bad" genes, those responsible for diseases and disabilities, with "good" genes.

MENDEL'S DREAMS—TODAY'S REALITY

These "impossible" feats have taken place in a science only about 100 years old. In 1865 Gregor Mendel, an Austrian monk, published the results of his 5-year study of inheritance in the garden pea plant. His work, which went unrecognized for 35 years, signaled the start of the modern science of genetics. He formulated basic laws of heredity, assuming the existence of a unit of heredity which he called "a factor" and which we call a gene. Since then, not only have genes and chromosomes been discovered but single genes have been isolated and synthesized. Mendel had a dream about genes that came true.

ISOLATING GENES

The first genes to be isolated, those that make ribosomal RNA (rRNA) in two species of African claw toads, were isolated in 1967 by Max Bernstein and Hugh Wallace at the University of Edinburgh, Scotland. At the same time D. E. Kohne of the Carnegie Institute, Washington, D. C., picked out the gene in *E. coli* that also transcribes rRNA.

A year later came the announcement that Jonathan Beckwith and his team of Harvard researchers had "snipped" out of the *E. coli* chromosome the lactose operon, the theoretical construct visualized by Jacob and Monod. This gene was isolated with the help of a virus which can detach a small piece of DNA from the *E. coli* chromosome. The virus cut out the entire lactose operon with its operator and promotor genes. Under the electron microscope the gene looks like any other piece of DNA—a twisted double stranded thread about 5000Å or one 50-thousandth of an inch long.

THE KHORANA GENE

A few months later came another "impossible" mission accomplished, a gene made in a test tube that works in a cell. After 9 years of research, H. Gobind Khorana and a team at the Massachusetts Institute of Technology made biochemical history and provided the ultimate proof that genes, pieces of DNA, are the genetic basis of life. Khorana developed techniques for joining short pieces of DNA of known base composition. Using the expertise he developed, he synthesized the gene that produces tyrosine transfer RNA molecules, modeled after the real molecule present in the chromosome of *E. coli.* The tRNA molecule made by this gene binds only to the amino acid tyrosine and transfers it to the tyrosine codon RNA in the ribosome. His first attempt to make a gene, in 1970, resulted in the creation of a "dead gene" modeled after the alanine tRNA gene found in yeast. He selected that gene because the entire base sequence of this transfer RNA molecule had been determined by Robert W. Holley. The Khorana team, then at the University of Wisconsin, made an exact duplicate of the natural alanine transfer RNA gene and put it into a yeast cell to test it. It failed to work because it lacked "stop" and "go" signals, without which a gene is "dead."

The second Khorana gene for tyrosine tRNA included the "stop" and "go" signals, the promotor and terminator genes. The completed gene was then introduced into *E. coli* where it functioned normally and produced tyrosine tRNA molecules. The implications of this achievement in genetic engineering could be heard loud and clear.

THE INNER LIFE OF A VIRUS

Another fantastic "first" in 1976 was the disclosure of the gene makeup of an entire organism, of one of the smallest and simplest of the viruses. A Belgian team led by Walter Fiers at the University of Ghent completed its study of the genetic composition of MS2, a virus that lives in *E. coli.* MS means "male specific", since this virus lives only in male bacteria—which it was previously noted, can mate. Genes pass from the donor or male across a bridge into the recipient or female bacterium.

MS2 is a spherical RNA virus, about 200Å or one millionth of an inch in diameter. It contains a single, circular thread of RNA composed of 3569 bases. The chromosome carries information for three genes. The first gene is coded for protein A, an enzyme that enables the virus to get into its host, *E. coli.* The second is the coat gene coded for the protein that makes a coat around the virus. The third is the replicase (REP-lick-ace) gene which enables the virus to replicate, that is, reproduce, inside its host. MS2 can make 10,000 copies of itself in about 15 minutes with these three genes, a rate of 10 viruses per second.

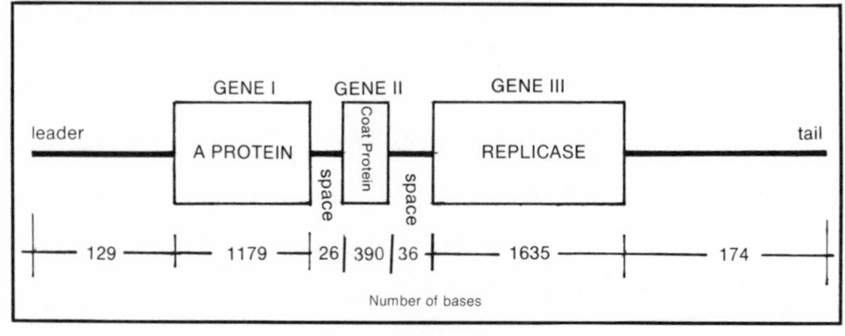

Chromosome of the MS 2 virus. It is a single thread of RNA composed of 3569 bases which carries information for three genes.

The gene makeup of MS2 was ascertained by breaking the chromosomes into smaller and smaller pieces by means of enzymes and separating the fragments. The base sequence of each of the smallest pieces was determined and then the pieces were assembled like a jigsaw puzzle.

The chromosome, it was disclosed, consists of a "leader" made up of 129 bases, followed by protein A of 1179 bases, a space of 26 bases, the coat protein of 390 bases, a space of 36, the replicase gene of 1635 bases, and a tail segment of 174 bases. The three genes appear to be regulated by a feedback mechanism. The replicase gene is controlled by the coat gene and the coat gene represses the replicase gene. When the coat gene is not coding, it folds up into many turns and loops like the petals of a flower. Other parts of the chromosome fold similarly and the entire chromosome resembles a bouquet.

THE DNA SURPRISE PACKAGE

Another first, with unexpected side effects, came in 1977 with the complete sequencing of another DNA virus—Phi X 174 (phi stands for bacteriophage and the number is its social security number for identification). The conqueror is Fred Sanger of insulin fame and his coworkers at Cambridge University. They reported the exact sequence of the 5375 bases making up the nine genes in Phi X 174. It is among the smallest of the *E. coli* phages, about 250Å or one millionth of an inch in diameter, containing a single strand of DNA about 18Å long. This would appear to be an exception to the Crick-Watson double-helix model but it was found that upon entering *E. coli* this phage makes another thread and becomes a double-threaded genetic conformist. The entire analysis was completed in 2 years, a new science track record for the event, thanks to the improved base sequencing techniques.

What is unusual about Phi X 174 is the great amount of genetic information compressed into its small chromosome. In fact, more proteins were being produced by this virus than could be accounted for by its genetic content. It seemed to contradict one of the basic tenets of modern genetics, "one gene, one protein"—that is, each gene produces only one protein. The mystery of the hidden genes was solved when it was discovered that there are no extra genes—the virus has an unusual way of reading the genetic information that it contains. Up to this point, genetic messages were read as you read this sentence, word after word in one direction with no overlap. A code word is read only once and the number and kind of amino acids in the protein correspond to the number and kind of code words (three-base codons) in the sentence (the gene).

Overlap of genes in Phi X 174 virus.

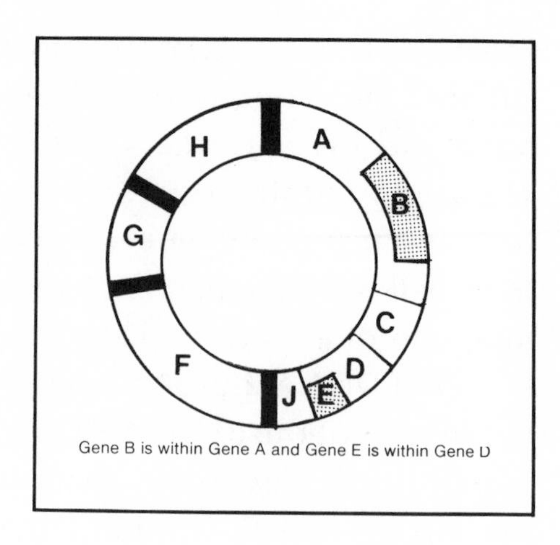

THE VIRUS READING METHOD

Phi X 174's secret is to have read the same message twice, starting at different places in the sentence, so that two messages are read and two proteins are produced. This can be illustrated by reading a sentence consisting of three-letter words twice. In the first reading, start with the first letter, C in the sentence that follows. In the second reading begin with the second letter, A.

C A N D O N E A T E E L

What is the first message? What is the second? Make up your own double message.

Overlapping, that is, the gene-within-a-gene situation, was discovered in the genetic instructions for two Phi X 174 proteins. Gene D is coded to produce a protein needed to make new viruses. Inside this gene is another gene, E, which produces another protein, an enzyme that helps destroy the host cell when the newly created viruses are ready to leave. Gene E starts in the middle of gene D but terminates just before the end of gene D. Gene E is contained entirely within gene D and the two have entirely different base sequences. By reading the message twice, starting at different points, there are two different genetic messages and therefore two different proteins are synthesized.

The second place where there is a shift in the reading frame is in gene A, which is housed entirely within gene B. Gene A codes for an enzyme which nicks the chromosome in starting DNA duplication; gene B provides the protein involved in making new viruses.

Sharing or overlapping is also found in the start-stop signals located between neighboring genes. The third base in the codon that signals the end of gene D is part of the starting signal of gene J.

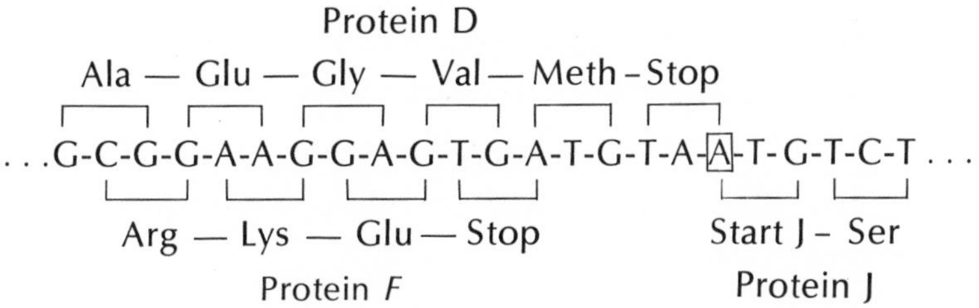

```
                 Protein D
  Ala — Glu — Gly — Val — Meth - Stop
...G-C-G-G-A-A-G-G-A-G-T-G-A-T-G-T-A-[A]-T-G-T-C-T...
    Arg — Lys — Glu — Stop          Start J - Ser
          Protein F                   Protein J
```

Similarly, one base of the stop signal of gene A is part of the start signal of gene C.

These revelations explain how tiny viruses are able to get the most out of their genetic material. This raises questions about coding in general: Is overlapping limited to bacteria and viruses or does it also occur in higher organisms? Do we have to reexamine our notions about the coding potential of genes? Also, now that we have disclosed the genetic makeup of a simple organism with 5000 or so bases, how long will it be before the chromosomes of higher organisms with a million or so bases can be analyzed?

SUMMARY

We have moved from the era of making models which describe the structure and functions of DNA to the era of making the genes themselves. The age of genetic engineering is dawning. Genes such as the lactose operon, once a theoretical idea, have been isolated and have become the models for synthesis. An artificial gene, a replica of a real tyrosine transfer RNA gene, has been synthesized and it works. The complete genetic makeup of two organisms is known—MS2 and Phi X 174. Another way of reading the genetic message has been revealed which makes it possible for a single gene to code for more than one protein, a revolutionary discovery in genetics.

14

THE GREAT DNA DEBATE

Is DNA research on the verge of creating a Frankenstein monster or a new and better world? Will scientific weapons kill us or cure us? Where is DNA research going? Are safeguards needed to direct and control DNA research?

These are some of the questions being debated by scientists and nonscientists around the world. New words and phrases fill the air and the news media—"recombinant DNA," "genetic engineering," "human engineering," "gene grafting," "gene splicing," "gene therapy," and "cloning." These terms are bandied about, causing consternation and confusion. The issue referred to as "recombinant DNA" is being discussed and debated on college campuses, in city councils, state legislatures, and congressional committees, in public forums, at international conferences here and in many other countries. The issue is expected to be with us for some time to come. Like it or not, we are going to have to face the problems and challenges being generated by DNA research all over the world. There is no stopping it or going back. It is part and parcel of our world today and tomorrow. Part of the problem is the lack of understanding of the issues by the nonscientific community and the lack of agreement in the science community. Ignorance begets fear and fear begets confusion, indecision, and indifference. Closing the information gap is one way of clarifying the issues surrounding the great gene-splicing debate. We hope this chapter will provide the basis for thinking about and acting on forthcoming legislation designed to define the limits and conditions for conducting DNA research. It addresses itself to answering such pressing questions as: How did the present recombinant DNA controversy arise? What are the basic issues? What are the possible solutions?

DNA RESEARCHERS FEAR DNA RESEARCH RESULTS

The present dilemma was born in science laboratories around the world. It is a product of the fantastic progress being made in DNA-RNA research, which is moving at mind-boggling speed. Only a quarter of a century ago, Crick and Watson constructed their theoretical Tinkertoy model of DNA. Today biologists can construct a DNA molecule and add it to a living organism, thereby changing a few of its properties. This new and growing power has many social and scientific implications. The first ones to see possible dangers in recombinant DNA research were a few of the researchers themselves, some of whom are outstanding scientists, including Nobel prizewinners. They were afraid of the possible creation of an organism so changed that it would be dangerous and uncontrollable, with the potential of starting an epidemic. The techniques of recombinant DNA research provided the means for masterminding new and unheard of forms of current life whose behavior was not completely predictable. "Monsters" might escape from the laboratory and perhaps spread some dangerous disease. Fear of the possible creation of such "biological atomic bombs" haunted scientists.

Weighing the expected benefits against possible hazards led a few of the researchers to ask questions about the possible environmental consequences of their research. However, we must remember that the added DNA is only a little part of the host DNA and the resultant combination cannot be considered a new form of life.

MINI-DNA RINGS: PLASMIDS

The term *recombinant DNA* needs some explanation. It describes a technique which imitates a process that goes on naturally in all living things from microbes to man. As the result of mating, pieces of chromosomes pass from one cell into another, usually within the same species. More than 25 years ago, Joshua Lederberg, a Nobel prizewinner in 1958, and Norton D. Zinder, then at the University of Wisconsin, discovered that certain phages can pick up pieces of bacterial DNA and tranfer it to the next bacterium they invade, usually of the same species.

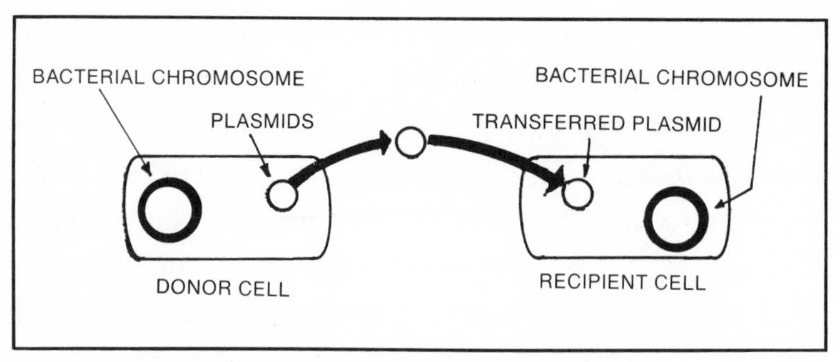

Plasmids transferred from bacterium to bacterium.

Today Zinder is working on new phages that tranport foreign genes, thus repeating in a test tube what he codiscovered a quarter of a century ago in living phages.

More recently another DNA transfer agent was discovered, plasmids—tiny doughnut-shaped rings of DNA found in almost all bacteria in addition to the chromosome. A bacterial cell may contain from 1 to 20 plasmids which are capable of reproducing independently of the host chromosome. Plasmids have been described as primitive bacteriophages that have not acquired the ability to live outside their host. And bacteriophages can be thought of as plasmids with the capacity to survive outside a cell.

GENE SPLICING

A plasmid sometimes picks up a piece of DNA from the chromosome of its own cell and deposits it in the cell of a related species of bacteria. This is rare in nature. In 1973 Stanley N. Cohn and Annie C. Y.

Chang at Stanford University and Herbert W. Boyer and Robert B. Helling at the University of California at San Francisco constructed a DNA molecule by splicing together parts of two plasmids found in *E. coli*. When this hybrid DNA molecule was inserted in *E. coli*, it duplicated and expressed the genetic information from both plasmid parents. Next, plasmid genes from a different species of bacteria were spliced into plasmids and introduced into *E. coli*, where they showed the traits that they had displayed in their original host. Using the splicing techniques, genes from a toad were successfully transferred to *E. coli*.

RECOMBINANT DNA DEFINED

The term *recombinant*, as it is used in the present controversy, refers to the techniques of transferring foreign DNA into bacteria. Genes from rabbits, fruit flies, silk moths, toads, yeast, sea urchins, rats, mice, and bacteria have been cut out and spliced into plasmids and so introduced into a bacterial cell, usually *E. coli*. Some of these "put-together" plasmids, by virtue of their powers of replication, are present in multiple copies within the host cell, as many as 20. When the host chromosomes replicate, so do the "put together" plasmids, including the piece of splice-in foreign DNA. Within a few hours there are millions of bacterial and millions of copies of the "put together" plasmid doing what any ordinary plasmid does plus what the foreign genes "tell" them to do.

ENGINEERING RECOMBINANTS

Constructing a recombinant DNA molecule sounds very simple. It involves transplanting a bit of plant or animal DNA into a bacterium and producing "mythological" molecules which may have existed previously only in the imagination of their creator or may be beyond prior imagination. The technique has been developing slowly over the past quarter century. The real and unsung heroes of the recombinant DNA story are an army of newly discovered enzymes which control and direct every step in this process. They quickly and efficiently replicate, repair, join, cleave, and transcribe DNA molecules. The most astonishing are the enzymes that cleave the DNA into specific pieces and offer a new means for isolating specific fragments of DNA. These are the restriction enzymes, which were first isolated and reported in 1970 by Hamilton O. Smith* and K. W. Wilcox at Johns Hopkins University, Baltimore, Maryland. Since then over 80 kinds of "cutting" enzymes have been isolated from various kinds of bacteria. Some of these enzymes cut a double strand of DNA into fragments that have single-stranded tails at either end with the same number and same kind of bases. The single-stranded ends contain complementary bases which are "sticky" and can join together to form a closed ring. Fragments from different organisms can be spliced together to make a recombinant DNA molecule. After being introduced into the host cell by a virus or plasmid, they

*One of three recipients of the 1978 Nobel Prize in medicine or physiology for enzyme genetic research.

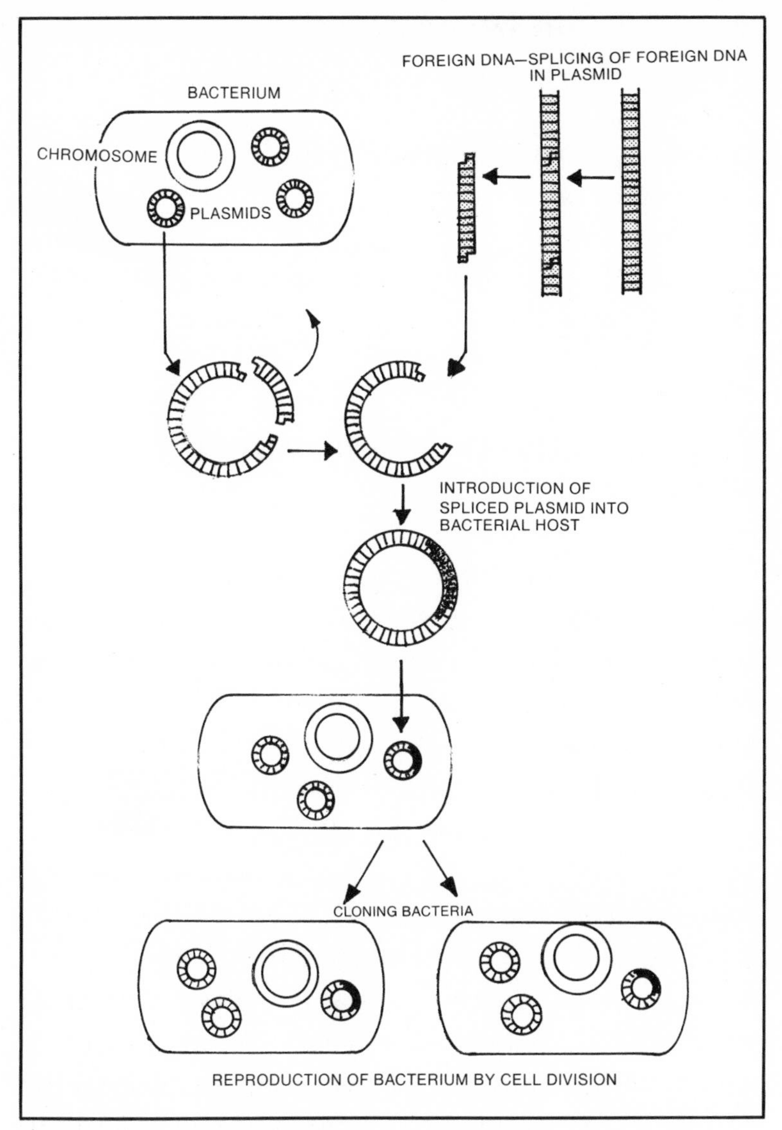

Recombinant DNA technique.

reproduce, mothering a group of organisms which contain identical gene combinations. The process of producing offspring with the same gene makeup is called cloning. It is a common method of asexual reproduction practiced by simple organisms bacteria, which usually reproduce by splitting in half. Propagating geraniums by stem cuttings or African violets by leaf cuttings are other examples of cloning. In recombinant research cloning is employed to produce large numbers of hybrid genes for study.

PROGRESS REPORT ON RECOMBINANT RESEARCH

A question frequently asked is: What is so special about recombinant research? Scientists cite many examples of research made possible by the recombinant techniques. With this new technique the base sequences of increasingly larger DNA molecules are being determined more quickly, more accurately, and more directly than before. Fred Sanger's analysis of the Phi X 174 virus with its 5375 bases took less than 2 years. C. Nigel Godson of Yale working with John C. Fiddes, one of Sanger's coworkers, hopes to complete the analysis of another virus, G4, in less than a year. These previously almost impossible tasks have been made feasible thanks to the rapid sequencing and splicing methods of recombinant research.

For the first time, it is possible to isolate specific segments of DNA from the complex chromosomes of higher animals such as fruit flies, toads, rabbits, and rats and splice them into carriers such as plasmids and viruses. These genes are transferred into host bacteria where they are reproduced in sufficient number to study their base sequence and gene action. By combining recombinant DNA and sequencing techniques, it is possible to isolate and study the genes of higher organisms. Now any DNA molecule that can be spliced into a virus or a plasmid can be sequenced. This is a giant step forward in DNA research and opens the way to understanding more about human genetics. Rabbit genes for making part of the hemoglobin molecule have been successfully inserted in *E. coli*, and genes for synthesizing insulin have also been transferred to *E. coli*. In both instances the foreign genes are alive and well but they have not yet been "turned on." Judging by past experiments, this is something to look for in the near future. A limitless supply of insulin synthesized by bacteria may be available in the next few years.

RECOMBINANT DNA RESEARCH RULES AND REGULATIONS

In June 1976, after several years of bitter debate, scores of private and public meetings and conferences, and public forums between scientists and nonscientists, the National Institute of Health (NIH), at the request of the National Academy of Science, issued a set of guidelines for recombinant DNA research. What is unusual about these rules and regulations is that they were formulated by the scientists themselves—because they were concerned with the possible social consequences of their research they voluntarily restricted the kinds of experiments that should be conducted. They also asked that an agency enforce these regulations. Despite differences of opinion among the scientists, they agreed that recombinant research should be continued. However, they proposed that precautions should be taken matched to the potential danger of each experiment. They also agreed that certain types of experiments should be forbidden because of their high potential risk. The guidelines are intended to safeguard the researchers themselves, the general public, and the environment.

A summary of the NIH guidelines for recombinant DNA research is as follows:

Five kinds of experiments are not to be started: cloning from disease-causing microbes or from cancer viruses; recombinants with genes for making powerful toxins; recombinants that might increase the strength and the range of plant disease genes; transferring genes for drug resistance to organisms that do not acquire it naturally; and deliberately releasing into the environment host organisms that contain recombinant genes.

Other experiments slightly less hazardous should be conducted under maximum security conditions guaranteed to prevent strange and unusual organisms from escaping into the environment. In general for experiments that can be done, four levels of security are described ranging from ordinary laboratory conditions found in most science laboratories to special laboratories previously required only where the most dangerous microbes were handled. Categories of host organisms incapable or unlikely to survive outside the laboratory were established.

These research rules and regulations are based on what might happen in the future, not on what has happened or is happening. They are precautionary measures. Concern for researchers, the public, and the environment is not new in science. Guidelines for handling radioactive materials, dangerous chemicals, and dangerous germs have been available and implemented for some time.

DNA DISARMED AND DEFANGED

One of the great fears of some gene-splicing biologists and the public is the DNA "devil" that escapes from the laboratory, accidentally or deliberately, runs amok and wreaks havoc. To minimize effects of such possible accidents, a new breed of "fail safe" organisms was engineered that could live only in the comforts of the laboratory and on a test-tube diet. These hothouse creatures are so domesticated that their chance of survival outside of a test tube is practically zero. Applying the techniques of recombinant experimentation to meet NIH ground rules, biological containment is added to physical containment as double security. The organisms playing the major role in these experiments are *E. coli* and its various phages and plasmids. They have been the chief targets in reconstruction efforts. For more than half a century a strain of *E. coli*, K-12, has been the darling of bacteriologists. It was adopted by the geneticists in the 1940s to replace the fruit fly and was accepted by the molecular biologists because more was known about it than any other living thing.

In January 1976 the first fail-safe edition of *E. coli* meeting the NIH safety rules was bioengineered by Roy Curtiss III and his team at the University of Alabama. They christened their masterpiece χ 1776, probably as a patriotic gesture. (The Greek letter is read as *chi,* pronounced kigh to ryhme with high.) This reconstituted version of *E. coli* K-12 is completely laboratory "broken." Its chances of survival outside the lab are about one in a billion. It is a pathetic shadow of its former self: it cannot live in the human gut or make its own cell wall, or exchange genes with other organisms nor be invaded by viruses. It does, however, have the blessing of NIH and is qualified to be used in recombinant experiments in combination with "pacified" plasmids and "vanquished" viruses. Herbert W. Boyer and Donald R. Helinski of the University at San Diego have constructed "pacified" plasmids. Safer bacteriophage carriers of DNA have been designed independently by Fredrick R. Blattner of the University of Wisconsin, Philip Leder of NIH, and Philip Sharp of MIT, and their coworkers. Some of these host-carrier combinations have NIH approval.

THE CLONING CONTROVERSY

Another issue that is attracting much attention and controversy is research related to genetic engineering through cloning. Cloning is a method of asexual reproduction in which a cell or a group of cells from the body of one parent organism produces an offspring genetically identical with itself. The nucleus in every cell carries all the genetic information required by an organism to reproduce itself. If one of these body cells can be induced to divide, it is possible for this one body cell from one parent to develop into an organism. This method of asexual reproduction, referred to as cloning, is found in nature among one-celled organisms, some insects, and plants; it is also used to propagate plants by stem and leaf cuttings, grafting and the like.

In the present controversy, however, cloning is defined as transplanting the nucleus from a donor cell into an unfertilized egg cell whose nucleus has been removed or destroyed. The resulting cell develops into an organism genetically identical with the donor. This technique has been carried out successfully with plants, fruit flies and frogs. Gordon, you will recall, produced a cloned frog by destroying the nucleus in an unfertilized frog egg cell and replaced it with the nucleus from an intestinal cell of another frog of the same species.

Although cloning has been successful in frogs, all known attempts in mammals, to date, have failed. It is theoretically possible to apply cloning to human beings. Experts in this field of research declare that many technical difficulties must be solved before human cloning is possible; this seems to be a long way off. Nevertheless, concerns about possible consequences of human cloning are being expressed.

Research using NIH-approved strains of *E. coli* in connection with recombinant DNA techniques has been highly successful in yielding significant scientific findings. In 1977 Howard M. Goodman and William J. Rutter of the University of California at San Francisco led a team of researchers who successfully inserted the insulin gene of a rat into the NIH-approved strain of *E. coli* with the help of recombinant bacterial plasmids. Many copies of the rat gene were produced in the host, which are useful in studying how this foreign gene works in a bacterium. However, the transplanted rat gene failed to make *E. coli* produce insulin.

Later in 1977, William Rutter coordinated three teams of California scientists from the University of California, the City of Hope Medical Center near Los Angeles, and the Salk Institute in La Jolla that were the first to induce rapidly growing bacteria to make a human hormone from a chemically synthesized gene. The hormone is somatostatin (so-mah-toe-STAH-tin), which is normally produced in the human brain and is believed to regulate the body's synthesis of insulin, growth hormone, and other substances. The hormone is important enough to have merited the 1977 Nobel prize in medicine or physiology for its discoverers, Roger Guillemin of the Salk Institute and Andrew V. Shalley of the Veterans Administration Hospital and Tulane University in New Orleans, Louisiana. The synthetic gene for the brain hormone was spliced into a plasmid and inserted into the NIH-approved strain of *E. coli*, where it directed the bacterium to make somatostatin, a chemical never before made by a bacterium. Finally in September 1978, scientists converted *E. coli* into human insulin factories. Synthetic genes for the A and B protein chains of insulin were inserted in *E. coli.* These chains were harvested, purified, and combined chemically into synthetic human insulin, the lastest recombinant DNA success.

FOR AND AGAINST RECOMBINANT RESEARCH

Despite the efforts of scientists to provide safeguards against the possible explosion of a "germ" bomb, the debate goes on. The basic issues at stake are expected benefits versus potential dangers on the one hand and control of recombinant DNA research on the other.

The supporters of recombinant research point to the burst of new knowledge gained by these techniques which may lead to such social and medical benefits as:

1. More effective, cheaper methods of producing many biological products—hormones, vitamins, antibodies, enzymes, antibiotics, and the like;

2. Increased food supply;
3. Better understanding of the causes of cancer;
4. New approaches to the energy problem; and
5. Increasing our knowledge and understanding of the living world around us, including ourselves.

The opponents of recombinant research see the possibilities of:

1. Worldwide epidemics of uncontrollable diseases;
2. Disastrous ecological disturbances;
3. A powerful tool for dictators, militarists, and terrorists; and
4. Scientists playing God, dominating and controlling all life including humanity.

THE DNA DILEMMA

Do the benefits of recombinant research outweigh the potential risks? There is no way of knowing—only time will tell. However, the most immediate question is a political one. By words and deeds, nonscientists are expressing the sentiment that recombinant research is far too important an issue to be left entirely to scientists. At present all federally funded recombinant research is governed by the NIH guidelines issued in 1976. It is expected that these guidelines will be extended by federal legislation to all such research in the private as well as the public domain.

Outside controls are being considered to monitor and evaluate the activities and the products of recombinant research. Shall they be local, state, national, or international controls? What limitations shall be set? By whom? These questions are difficult to answer and must wait until many of the problems and issues of science in its relationship to society are clarified and resolved.

How realistic is it to propose that DNA recombinant research in this country be constrained, restrained, or redirected in view of the international nature of science? A deluge of DNA data flows out of thousands of research laboratories around the world. The findings are published in hundreds of periodicals with worldwide circulation and presented at international conferences. There is a free flow of information and constant evaluation of scientific findings by other scientists for anybody who is interested. Researchers in practically all fields of science and in the social sciences read these publications, attend international conferences, and are discovering implications of DNA recombinant research in their own disciplines. To bind and gag the American researcher is hardly an effective solution considering the international character of recombinant DNA research.

On the other hand, American scientists as leaders in recombinant research should assume social responsibility for the products of their labors and also provide international leadership. The fears of some sectors of the scientific, public, and political communities—founded or unfounded, real or imaginary—are being voiced. Questions are being asked and answers must be provided. Once again it must be emphasized that the fears are based upon what may happen and not what has happened as a consequence of recombinant research. Worldwide epidemics that destroy life and land, dictators and terrorists who use DNA recombinants as a biological atomic bomb, and scientists who assume divine powers are all still in the realm of science fiction. The DNA recombinant restrainers argue that "it can happen here" and that "an ounce of prevention is worth a pound of cure."

There are signs that the tide of public opinion and political pondering may be changing toward recombinant DNA research in the United States. Consideration is being given for a modest relaxation of the NIH rules governing such research, as suggested by the NIH Recombinant DNA Advisory Committee. This action has been inspired, in part, by the recognition that the fail-safe of *E. coli* developed for these experiments is much safer than was originally believed. Also time has cooled tempers, clarified thinking, encouraged reflection, and has not been witness to an epidemic caused by a biological bomb or DNA demons. In general, DNA researchers have given a good account of themselves and their activities. This may have helped to restore public and political confidence in scientists' self-government and self-control about the impact of recombinant DNA research on the total environment.

THE DESTINY OF DNA

What is the future of recombinant DNA research? The DNA-RNA concept has profoundly altered our understanding of the nature of life on the molecular level. It has enabled us to gain new insights into the operation and interrelationship of mind and body. Hope for the conquest of hereditary and virus diseases as well as cancer grows stronger as recombinant research expands. The successful syntheses of genes and viruses are unprecedented landmarks in the history of man's efforts to control his environment. Humanity now has the basic know-how for changing itself and its world. As we climb the ladder of life, DNA research is opening up new vistas which point to the possibility of making this a better world in which to live.

The products and promises of the benefits of recombinant research cannot be dismissed or denied. As is so often true of scientific discoveries and inventions, the problem is the use to which they are put. There is every reason and hope to believe that whatever legislation is proposed for controlling recombinant DNA research will be designed to maximize its benefits and minimize its dangers.

GLOSSARY—DEFINITIONS OF IMPORTANT WORDS

Adenine (A). The purine base present in DNA and RNA.

Albino. An organism lacking colors normally present. Human albinos lack the pigment melanin, normally present in skin, hair, and eyes, because of a gene disorder; they have white skin, white hair, and pink eyes.

Amino Acids. The chemical molecules of which proteins are composed. There are 20 kinds of amino acids; they can be linked to form an infinite number of protein molecules. (Example: alanine, $C_3H_7NO_2$.)

Angstrom (Å). A unit of length in the metric system equal to one ten-billionth of a meter or one 250-millionth of an inch. It is used in measuring the length of light waves, atoms, molecules, and viruses.

Antibody. A protein produced in an organism in response to the presence of an antigen, that is, a foreign substance with which it interacts.

Anticodon. The triplet of bases (codon) on the loop of a transfer RNA molecule that matches the codon of messenger RNA.

Asbestos. A fibrous mineral which does not burn and conducts heat very slowly. It has carcinogenic properties, causing lung cancer in humans.

Atomic Energy. The energy released by fission of fusion of the nucleus of an atom in the form of radiation or high-energy particles.

Atoms. The smallest unit into which a chemical element can be divided and still retain its characteristics.

Bacteriophage (phage). A virus that lives and reproduces inside a bacterium. Phi X 174 is a phage for the bacterium *E. coli.*

Bacterium. A microscopic organism with a cell wall and a chromosome not enclosed in a nuclear membrane. (Example: *E. coli.*)

Base. The purine and pyrimidine molecules present in DNA and RNA: adenine, guanine, thymine, cytosine, and uracil.

Base Pairing. The process whereby adenine (A) always pairs with thymine (T) or uracil (U), and guanine (G) with cytosine (C), to form a double-threaded nucleic acid molecule.

C^{14}. A radioactive form of carbon that emits weak rays; its half life is 5700 years.

Cancer. A group of diseases characterized by uncontrolled and abnormal cell growth.

Carcinogens. Chemical, physical, or biological agents that cause cancer.

Cell. A microscopic unit of structure and function of living things. A mass of living matter surrounded by a membrane.

Cell Culture. The technique of growing cells isolated from many-celled organisms.

Cell Division. The dividing of a parent cell into two daughter cells.

Cell Membrane. The outer surface of a cell. It regulates the materials that enter and leave the cell.

Central Dogma. The concept that DNA is the template for its own duplication and for RNA. RNA, in turn, is the template for protein synthesis. DNA → RNA → proteins.

Chromatin. The nucleoprotein fibers of which chromosomes are composed in cells with a nuclear membrane.

Chromatography. The technique of separating different kinds of molecules by permitting a solution of the mixture to migrate along a strip of filter paper or a column of other absorbent material.

Chromosomes. The threadlike structures, either linear or circular, in a cell which contain hereditary material (DNA or RNA).

Clone. A line of cells or cell structures all descendants from a common ancestor and therefore genetically identical.

Codon. The sequence of three bases that code for a particular amino acid.

Color Blindness. A human hereditary disease in which the sufferer cannot distinguish between red and green colors. It is more common in males.

Complementary Base Sequence. Two nucleic acid threads joined by base pairing (A-T, C-G, or A-U).

Compound. A substance composed of two or more chemical elements combined in definite and constant proportions.

Cosmic Rays. Powerful rays from outer space with great penetrating ability. They may be the cause of spontaneous mutations.

Cytoplasm. The living material in a cell between the cell membrane and the nuclear membrane.

Cytosine (C). The pyrimidine base found in DNA and RNA.

DDT (*d*ichloro-*d*iphenyl-*t*richloro-ethane). A synthetic insecticide once used to control insects that carry disease and destroy crops; today it is banned for almost all use because of its harmful effects.

Delaney Clause. A provision of the 1958 amendment to the Pure Food, Drug, and Cosmetic Act that bans the use in foods and beverages of any food additive that is shown to cause cancer in humans or animals.

Deletion. The loss of a segment of a chromosome, which may include one or many genes.

Deoxyribose. A five-carbon sugar present in DNA.

DES (*di*ethyl*s*tilbestrol). A synthetic female sex hormone. It is linked to cancer in the daughters of mothers treated with this hormone during their pregnancy.

DNA (*d*eoxyribo*n*ucleic *a*cid). The molecule which contains the genetic material in a cell.

DNA Polymerases. The enzymes involved in DNA duplication. Polymerase I, the first one discovered, seems to be involved in DNA repair. Polymerase III, the most recent discovery, seems to be concerned with DNA duplication.

Double Helix. The Crick-Watson model of DNA, consisting of two DNA threads arranged in a spiral.

Electron. A negatively charged particle that moves around the nucleus of an atom.

Electron Microscope. An instrument for magnifying objects using a stream of electrons instead of light rays. Magnification and resolution attainable are a thousand times greater than with a light microscope.

Element. A substance containing only one kind of atom. It is one of the 105 natural or man-made kinds of atoms which alone or in combination with other elements make up all the matter in the universe. (Example: carbon.)

Endocrine Glands. Glands of the body that secrete their products—hormones—directly into the bloodstream.

Endoplasmic Reticulum. The network of double membranes in the cytoplasm of a cell. Often coated with ribosomes.

Environment. All the factors that surround and affect a living thing.

Enzymes. Protein molecules that control the chemical reactions in an organism.

Evolution. The changes in the genetic makeup from one generation to the next. It is the result of natural selection acting on random genetic variations.

Fallout. The slow descent of radioactive particles from the atmosphere after an atomic bomb explosion.

Feedback. A self-regulating mechanism in living things. Part of the output of the system is fed back into the system to regulate further output. (Example: the thermostat in a heating system.)

Fertilization. The fusion of two sex cells, the sperm and the egg, to produce a fertilized egg capable of growing into a new organism: this is sexual reproduction.

Fission. A fast method of asexual reproduction in which a one-celled organism splits into two daughter cells, genetically identical with the parent.

Fossil Fuels. The remains of living things burned to obtain energy. (Examples: coal, oil, and natural gas.)

Gametes. Sex cells that unite at fertilization; eggs and sperm.

Gene. A part of the hereditary material (DNA or RNA) on a chromosome. It is coded for the synthesis of a specific protein.

Gene Splicing. The technique of joining together genes from different sources.

Genetic Code. The sequence of bases along a DNA (or RNA) molecule that determines the sequence of amino acids in the protein chain. The code consists of a sequence of three bases which is specific for an amino acid.

Genetic Dictionary. The codons for each of the 20 amino acids, as well as the "stop" and "start" signals.

Glucose. A six-carbon sugar used as an energy source in practically all living things.

Guanine (G). A purine base present in DNA and RNA.

H^3 (tritium). A radioactive form of hydrogen which gives off weak radiations; it has a half life of 12.5 years. Used to tag molecules.

Hairpin Loops. Portions of single DNA or RNA strands formed by the pairing of complementary bases and resembling the double helix.

Half-Life. The length of time required for half the atoms in a radioactive material to lose their ability to give off radiations.

Helix. A spiral structure with repeating units.

Hemoglobin. An iron containing protein found in red blood cells. It carries oxygen and carbon dioxide.

Hemophilia. A hereditary human disease in which the blood does not clot, causing excessive bleeding from small wounds.

Hereditary Diseases. Disease caused by DNA disorders; a gene mutation, which is passed on from parent to offspring.

Heredity. The transfer of characteristics from parent to offspring by the transmission of genetic material (DNA or RNA).

Herpes. A virus with a linear double-helix DNA chromosome with possible carcinogenic properties.

Histone. A protein found combined with DNA in cells with a nuclear membrane.

Hormone. A chemical substance secreted directly into the bloodstream by an endocrine gland. It produces specific effects in other parts of the organism.

Host Cell. A cell used by viruses for growth and reproduction; a cell on or in which a parasite lives.

Hybrid. An offspring that differs from its parent in one or more traits.

Hydrogen Bond. A weak chemical bond that links a hydrogen atom to another atom.

Immunoglobulins (gamma globulins). Y-shaped protein molecules which combine with foreign chemicals.

Immunology. The study of antibodies and their reactions to foreign chemicals (antigens).

Inducers. Molecules that stimulate the production of large amounts of enzymes. They are part of the gene control mechanism in microbes.

Initiation Factors. Proteins needed to start protein synthesis.

Insulin. A hormone, produced by the pancreas, that enables cells to use glucose.

Lampbrush Chromosomes. Giant chromosomes with loops projecting in pairs. The loops are sites of DNA activity.

Larva. An immature stage in the life history of an animal; the caterpillar in insects and the tadpole in frogs.

Leukemia. A form of cancer of the blood-forming tissues. It involves a great growth of abnormal white blood cells.

Lysozymes. Enzymes that dissolve the cell walls of certain kinds of bacteria.

Melanin. A group of dark brown or black pigments present in the skin, hair, and eyes. Albinos lack melanin.

Messenger RNA (mRNA). RNA molecules, made by DNA,

that contain genetic information for protein synthesis which is carried to the ribosome where it determines the order of the amino acids in the protein molecules made.

Metabolism. The sum of all the chemical and physical processes within a living thing.

Micrometer (μm). A unit of length in the metric system equal to one millionth of a meter or one 25-thousandth of an inch. Commonly used as a unit of measure of cells. It was formerly called the micron (μ).

Mitochondria. Structures in the cytoplasm of a cell which carry on oxidation.

Molecule. The smallest unit of a substance that consists of two or more atoms.

Mutagen. A chemical, physical, or biological agent that increases the rate of mutations.

Mutant. A changed gene or an organism containing a changed gene.

Mutation. An inherited change in a chromosome due to an alteration in its DNA or RNA.

Myoglobin. A protein present in muscles. It is similar to hemoglobin in the blood in structure and function.

N^{15}. Heavy nitrogen, and isotope of N^{14}, the common form of this element.

Neutron. An uncharged particle found in the nucleus of all atoms except hydrogen (H).

Nitrous Acid. (HNO_2). A very powerful mutagen capable of altering the chemical composition of DNA bases.

Nonhistones. Proteins associated with chromosomes.

Nucleic Acid. A polymer consisting of sugar, phosphate, and four different bases (from the group A, T, C, G, and U). (See DNA; RNA.)

Nucleolus. A round, granular structure present in the nucleus of a cell. It contains DNA, RNA, and proteins, and participates in the synthesis of rRNA and ribosomes.

Nuclear Membrane. A double membrane with pores that separates the nucleus from the cytoplasm.

Nucleus (of an atom). The center of an atom. It is composed of protons and neutrons in all atoms except hydrogen (H).

Nucleus (of a cell). The structure that contains genetic information in the form of DNA or RNA. It is enclosed by a double membrane in almost all cells except those of bacteria, viruses, and simple algae.

Operator. A portion of the DNA molecule capable of reacting with a specific repressor and thereby controlling the activities of an operon.

Operon. A group of neighboring genes that operate together under the control of an operator and a repressor gene.

Organ. A part of the body composed of several kinds of tissues working together as a structural unit. (Examples: kidney, stomach.)

Organic Chemicals. Compounds formed by living things; carbon compounds such as sugar, fats and proteins, DNA and RNA.

Organism. A living thing composed of one or of many cells. (Examples: a bacterium, a human.)

P^{32}. A radioactive form of phosphorus with a half life of 14.3 days. It gives off powerful particles.

Pancreas. An organ in the body of humans and other animals that secretes the hormone insulin into the bloodstream.

Parasite. An organism that gets its food from another living thing by living either on or in its host. A bacteriophage eats its host, a bacterium.

Phage. See bacteriophage.

Pituitary Gland. An endocrine gland located at the base of the brain in humans. It secretes several hormones including the growth hormone.

Plasma Membrane. See cell membrane.

Plasmids. Circular, independently reproducing chromosomes present in the cytoplasm of bacteria.

Pollution. The degrading of the quality of the environment by introducing impurities harmful to living things.

Polymer. A large molecule composed of many repetitive smaller molecules. (Examples: DNA, RNA.)

Polymerase. See DNA polymerases; RNA polymerases.

Promotor. The portion of the DNA molecule to which RNA polymerase binds and starts transcription.

Proofreading. The process by which incorrect bases are removed and replaced by the correct ones.

Puffs. Open loops on chromosomes similar to those present on lampbrush chromosomes.

Purines. A group of nitrogen-containing compounds with two rings, such as adenine (A) and guanine (G), present in DNA and RNA.

Pyrimidines. A group of single-ringed, nitrogen-containing compounds, such as cytosine (C), thymine (T), and uracil (U), present in RNA.

Radioactivity. The emission of radiation by the nucleus of an atom such as uranium.

Reading Mistake. The incorrect placement of an amino acid during protein synthesis, as in sickle-cell anemia.

Regulatory Genes. Genes that control the rate at which other genes produce a substance.

Repair Synthesis. The removal and replacement of damaged portions of DNA by enzymes.

Replicating Fork. Y-shaped portion of the chromosome where DNA duplication is in process.

Repressors. A special group of protein molecules that control the rate at which mRNA is made.

Resolving Power. The ability of a microscope to distinguish two lines separately. The resolving power of the best light microscope is 0.2 micrometer; with the electron microscope this has been increased a thousand times to 2 angstroms.

Respiration. The process of capturing the energy of glucose and other fuel foods in a form that can be used by cells.

Restriction Enzymes. Enzymes that cut the double-threaded DNA molecule at specific places.

Reverse Transcriptase. An enzyme produced by certain RNA viruses that makes single-stranded DNA chains from RNA templates. RNA→DNA.

Ribonuclease. An enzyme that splits the RNA molecule.

Ribosomal Proteins. Special proteins found in the ribosome bound to rRNA.

Ribosomal RNA (rRNA). One kind of RNA molecule present in the ribosome. Its function is unknown, but it has some role in protein synthesis.

Ribosomes. Small particles where protein synthesis takes place. They consist of rRNA and proteins, and are found either free in the cytoplasm or attached to the outer membrane of the endoplasmic reticulum.

RNA (Ribonucleic acid). A single-stranded molecule of nucleic acid containing repeating bases: adenine (A), guanine (G), cytosine (C), and uracil (U).

RNA Polymerases. Enzymes that are involved in the synthesis of RNA on a DNA template.

S^{35}. A radioactive form of sulfur. It has a half-life of 87 days.

Scanning Electron Microscope (SEM). An electron microscope that produces a three-dimensional image of surfaces of an object.

Sickle-Cell Anemia. A hereditary disease which results in anemia because the hemoglobin molecules are defective.

Sigma Factor. The part of the RNA polymerase molecule that recognizes the specific place on DNA for starting RNA synthesis.

Spontaneous Mutations. Mutations for which there is no apparent cause.

"Sticky" Ends. Single-stranded tails projecting from the ends of double-threaded DNA molecules in which the bases are complementary.

Structural Genes. The genes in an operon which produce mRNA for making a protein. Their functron is controlled by the operator and regulator genes.

Template. A mold or pattern for making another molecule. DNA is the template for the synthesis of DNA and also of RNA.

Termination Codon. Codon that signifies the end of the amino acid chain synthesis.

Thymine (T). A pyrimidine base present in DNA.

Thymus. An endocrine gland located in the upper chest cavity. It is important in the development and maintenance of immunity.

Thyroid Gland. An endocrine gland located in the neck on either side of the voice box. It controls the rate of metabolism in the body.

TMV (tobacco mosaic virus). A virus that causes a disease in the tobacco plant. It is widely used in research.

Transcription. The process by which the sequence of the bases in DNA is used to order the base sequence in an RNA molecule.

Transfer RNA (tRNA). A type of RNA molecule that is coded to attach to a specific amino acid and place it in the correct position on mRNA. There are 20 kinds of tRNA, one for each kind of amino acid.

Translation. The process by which the genetic message in mRNA directs the sequence of amino acid molecules in protein synthesis in the ribosome.

Tumor. A mass formed by the uncontrolled growth of cells.

Ultraviolet Rays. Invisible rays that can induce mutations.

Uracil (U). A pyrimidine base present in RNA.

Virus. A disease-causing infectious agent consisting of a protein coat surrounding a nucleic acid core of either DNA or RNA. Viruses can reproduce only inside a host cell.

X-ray Crystallography. The technique of obtaining patterns produced by X rays scattering from crystals to determine the three-dimensional structure of molecules.

SUGGESTIONS FOR FURTHER READING

Butler, J. A. V. *Gene Control in the Living Cell.* New York: Basic Books, 1968.

Clarke, Robert F. *DNA, Action Model.* Minneapolis: Burgess Publishing Co., 1968.

Curtis, Helena. *The Viruses.* Garden City, N.Y.: Natural History Press, 1965.

Dubos, René. *The Professor, The Institute, and DNA.* New York: Rockefeller University, 1976.

Engel, Leonard. *The New Genetics.* Garden City, N.Y.: Doubleday & Co., 1967.

Kendrew, John. *The Thread of Life.* Cambridge, Mass.: Harvard University Press, 1966.

Lear, John. *Recombinant DNA—The Untold Story.* New York: Crown Publishers, Inc., 1978.

Lessing, Lawrence. *DNA at the Core of Life Itself.* New York: Macmillan Publishing Co., 1967.

Luria, S. E. *Life in the Unfinished Experiment.* New York: Charles Scribner's Sons, 1973.

Olley, Robert. *The Path of the Double Helix.* Seattle: University of Washington Press, 1974.

Sayre, Ann. *Rosalind Franklin and DNA.* New York: W. W. Norton & Co., 1975.

Strickberger, M. W. *Genetics, 2d ed.* New York: Macmillan Publishing Co., 1976.

Sullivan, Navin. *The Messenger of the Genes.* New York: Basic Books, 1967.

Wade, Nicholas. *The Ultimate Experiment, Man-Made Evolution—DNA.* New York: Walker & Co., 1977.

Watson, James D. *Molecular Biology of the Gene, 3d ed.* Menlo Park, Calif.: W. A. Benjamin, 1976.

Winchester, A. M. *Genetics, 5th ed.* Boston: Houghton Mifflin Co., 1977.

Woodward, D. O. and Woodward, V. W. *Concepts of Molecular Genetics.* New York: McGraw-Hill, Inc., 1977.

INDEX